T0309615

Vision: Images, Signals
and Neural Networks

Models of Neural Processing in
Visual Perception

PROGRESS IN NEURAL PROCESSING*

Series Advisor
Alan Murray *(University of Edinburgh)*

*For the complete list of titles in this series, please write to the Publisher.

Progress in Neural Processing • 19

Vision: Images, Signals and Neural Networks

Models of Neural Processing in Visual Perception

Jeanny Hérault

University Joseph Fourier &
Institut National Polytechnique of Grenoble
France

W⊖ World Scientific

NEW JERSEY • LONDON • SINGAPORE • BEIJING • SHANGHAI • HONG KONG • TAIPEI • CHENNAI

Published by

World Scientific Publishing Co. Pte. Ltd.

5 Toh Tuck Link, Singapore 596224

USA office: 27 Warren Street, Suite 401-402, Hackensack, NJ 07601

UK office: 57 Shelton Street, Covent Garden, London WC2H 9HE

British Library Cataloguing-in-Publication Data
A catalogue record for this book is available from the British Library.

Progress in Neural Processing — Vol. 19
VISION: IMAGES, SIGNALS AND NEURAL NETWORKS
Models of Neural Processing in Visual Perception

Copyright © 2010 by World Scientific Publishing Co. Pte. Ltd.

ISBN-13 978-981-4273-68-8
ISBN-10 981-4273-68-6

Typeset by Stallion Press
Email: enquiries@stallionpress.com

Printed in Singapore by B & Jo Enterprise Pte Ltd

Dedicated to

my beloved wife, who endured with so much stoicism the writing of this book,

my colleagues and my students of TIRF, LIS and GIPSA laboratories, with whom we have had so many frank, friendly and fruitful discussions about vision, high-dimensional spaces, independent component analysis, and much more...

Bernard Ans
David Alleysson
William Beaudot
Gérard Bouvier
Alan Chauvin
Pierre Demartines
Barthélémy Durette
Giansalvo Cirricione
Anne Guérin-Dugué
Brice Chaix de Lavarenne

Aude Oliva
Boubakar Séré
Corentin Massot
Christian Jutten
Denis Pellerin
Hervé Le Borgne
Nathalie Guyader
Christian Marendaz
Antonio Torralba
Patricia Palagi-Monteiro

Preface

This book results from many years of research and teaching in a cognitive approach to vision modeling. It relates the point of view of an engineer, specialist of signal and image processing, who is willing to discover and understand the basic principles of the Human visual system, with the aim to design high-level image processing systems. In fact it is fundamentally multidisciplinary and it continuously refers to a "triangle" constituted of three main disciplines:

Neurobiology, Mathematics, Psychophysics.

At first sight, the first discipline — Neurobiology — is the very source of data about the structures and functions of the visual system's neural circuitry. Due to the fantastic advances in neurobiology during the past decades, it is now possible to have a rather precise idea of the substratum of many visual processes. But sometimes, data is missing or incomplete (experimentation at the cell or membrane level is very difficult) so that it becomes difficult to understand the mechanism underlying some functions.

At this stage, the second discipline — Mathematics — appears as considerably helpful, especially through the theory of signals and its applications to signal and image processing. The mathematics are used to modelize the neural circuits and to formalize them into their "system" aspect. The role of formalization is to represent systems by means of sets of equations that help to categorize them easily. In another way, mathematics are particularly useful to formalize hypotheses in order to explain some given behavior or to predict other ones not yet observed. It is important to consider that mathematics are two-fold: (1) equations that constitute a code of representation and (2) manipulation of equations, which is relevant to a specialist. The neophyte or non-mathematician should not be disheartened by the apparent complexity of the mathematical formulae in this book: he (or she) has just to learn and remember the names and shapes of some equations and does not need to be able to make by him- or herself the calculations that are exposed. The text will explain which important facts and consequences are to be remembered.

Often, there are not enough biological data to explain a given property or a given behavior. Without any structural basis, mathematics cannot be of any use (no equation can be designed). The help will come then from the third discipline — Psychophysics — where experimentations can be conducted on animals or humans. Due to the recent and continuous progresses in experimental and cognitive psychology, very accurate results can be obtained in order to ascertain various hypotheses. Input-output relationships can thus be established, and system theory can be used together with signal theory to modelize the functional aspects of the sub-system under consideration.

Under such a scheme, the cooperation between *biology*, *psychology* and *signal processing* appears to be of capital interest for cognitive research. Nowadays many researchers in the world, issued from one discipline, have made the effort to come to a second one, and sometimes to the third one. This gives rise to very efficient research teams with wide-minded members able to produce a number of consistent works and significant breakthroughs.

In addition, due to the cross-fertilization between disciplines resulting from this approach, each specialist will greatly increase his or her knowledge in his or her domain, and reciprocally will be able to cast a new light on the other domains.

This book is intended for students who are newcomers in the domain of vision, as well as to confirmed researchers who are willing to discover or rediscover the various aspects of vision from the point of view of modeling. Together, neurobiologists, psychophysicists, mathematicians and researchers in signal and image processing may find in the presented approach a series of ideas and hints to help them to go further in their own specialty.

Regarding the way of reading this book, there are several possibilities. Some may take it as a popularization work on visual processes, and skim through the book in order to just catch a bird's eyeview on visual structures and their principal functions. Others may be in search for a more in-depth analysis of the visual functions through their mathematical models. They will explore all formulae and equations in order to understand the mechanisms of the visual system and then gather hints for further developments.

This book begins with a chapter devoted to the physics of light and its alterations when traversing media and optics. It will give a global idea of the complexity of the tasks which are required to the visual system before extracting useful information.

The second chapter consists in an engineer's view over the architecture of the visual system. It prepares the reader to the style of approach developed in the sequel.

The model of the retina is exposed in the third chapter, which summarizes all the basics of the retinal circuitry and the fundamental role of the retina as an image pre-processor.

In the fourth chapter, the reader will discover the concept of "neuromorphic circuits" and its application to motion estimation. Though not proven to exist in the visual system, this function is of interest for image sequences analysis.

The object of chapter five is the processing of color in the retina. It illustrates the power of biological systems to code information in a highly economical and efficient form.

Chapter six presents all the advantages of irregular sampling and of non-linear processing in the retina, making it an optimal adaptive system to extract information from the visual scene.

In chapter seven, it will be found a model of the primary visual area V1 with the fantastic role of complex cells for scene analysis, leading to algorithms for categorization, monocular perspective estimation and attentional processes. As in the retina, adaptation processes are shown and may provide an explanation to well known visual after-effects. A model of the McCollough effect illustrates these principles.

Jeanny Hérault

Contents

Chapter 1

From Photons to Image Formation

In this section, we will briefly review the basic principles in order to understand what is light, and how it is affected by the crossing of various media, objects and optics before reaching the photoreceptors.

1.1. Light and Lighting

1.1.1. *Nature of Light*

1.1.1.1. *Basics*

Electromagnetic theory

According to the electromagnetic theory, light is considered as the conjugation of an electric field E and a magnetic field H perpendicular to E, which simultaneously vibrate at the same frequency ν and propagate at a velocity $v = 1/\sqrt{\varepsilon\mu}$, ε and μ being respectively the permittivity and the permeability of the propagating medium. The trihedron (E, H, v) is rectangular and direct (Figure 1.1). In vacuum, the velocity of light is $c = 299\ 792$ km/s. The frequencies for visible light are around 10^{15} Hz.

As E and H are linked, only one of them can be considered for basic purpose, so we will use E because of its responsibility in receptors' responses:

$$E = E_0 \cos\left(2\pi\nu\left(t - \frac{x}{v}\right)\right) = E_0 \cos\left(2\pi\left(\nu t - \frac{x}{\lambda}\right)\right), \quad (1.1)$$

with $\lambda = v/\nu$ being the wavelength. In the case of visible light, the wavelengths in vacuum range from 380 to 750 nm. Such a wave, with H proportional to E is said to be *rectilinearly polarized*.

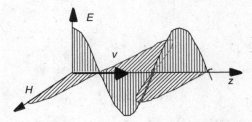

Figure 1.1. An electromagnetic wave: the electric and magnetic fields are orthogonal. The wave propagates at a velocity v in a direction perpendicular to the wave front.

Natural light

Natural light can be represented as composed of multiple rectilinear vibrations emitted by atoms or molecules and occurring at times when their excitation state varies. Such variations happen at irregular time instants and have durations of the order of 10^{-9} s, which is very long with respect to their period ($\approx 10^{-15}$ s) but very short with respect to the response time of receptors (≈ 1 ms). The vector E often varies in direction, amplitude and phase, and these variations are not perceived by the receptors. Its projection on a given axis O_x is perceived as being of constant intensity, the mean value of which being $E_0/2$, being independent of the orientation of O_x.

Polarization of light

In some natural cases (metallic or vitreous reflections, diffusion in gases ...), one direction of vibration is isolated, resulting in *polarized light*. For example, the diffusion of light by molecules of gas favors the vibration in a direction that is orthogonal to the plane containing the incident ray and the diffused ray. Polarization is complete when those rays are orthogonal: the blue light of the sky is completely polarized when observed in a direction which is normal to that of the sun.

Light carries energy

Light carries energy (linked to the amplitude of the vibrations) in the direction of propagation. However, the electromagnetic theory does not fully explain the interactions between radiation and matter: for this purpose, we need to consider light as composed of particles (photons), the energy of which is related to the frequency $W = h\nu$, h being the Planck's constant

(6.623 × 10^{-34} J s). One must note that the power of light depends on its frequency, or on its wavelength. For natural light which is composed of different multiple vibrations, the total power is the sum of all (wavelength) spectral components.

1.1.1.2. *Spectral Composition of Light*

The spectrum of electromagnetic radiation extends from Gamma and X rays to Ultra-Violet, Visible light, Infra-Red and Radio waves, according to Figure 1.2. From the point of view of energy, the higher the frequency (or the shorter the wavelength), the higher the energy.

All sources of light have an emission spectrum: they do not radiate uniformly over the wavelengths. Figure 1.3 gives some examples of sources like sun and black body.[1]

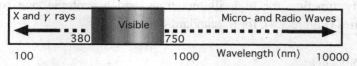

Figure 1.2. The spectrum of visible light between the ranges of gamma and X-rays on one side and the micro- and radio waves on the other side.

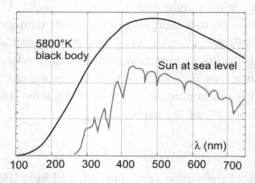

Figure 1.3. Shape of spectra: the 5800°K black body and the sun at sea level.

---

[1]The black body is an object that absorbs all light that hits it. When it is hot, it is an ideal source of thermal radiation.

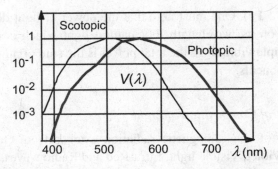

Figure 1.4. The visibility function V_λ. There are two definitions: one for dim light (scotopic) and one for daylight (photopic).

Visible light

The sensitivity curve is called "the visibility function" $V(\lambda)$ or V_λ. The Commission Internationale de l'Eclairage (CIE) has defined a photopic visibility function for high illuminations and a scotopic one for low intensities (Figure 1.4), for a "standard observer".

Ultra-Violet light

In Ultra-Violet, the UV-A radiations ($\lambda \sim$ 315–400 nm) are the least harmful and are used in phototherapy or as "black light" (causing fluorescent materials to emit visible light and glow in the dark). UV-B radiations (280–315 nm) have enough energy to cause severe damages in biological tissues. Solar UV-B are not completely absorbed by the atmosphere and necessitate some artificial protection (solar cream, solar spectacles). A natural protection exists for the eye's photoreceptors, due to their partial absorption in vitreous humor. Solar UV-C radiations (180–280 nm) are completely absorbed by the atmosphere.

Infra-Red light

On the other side of the visible spectrum, infrared light (IR) is essentially perceived through its thermal effects. Because of its low amount of energy per photon, it hardly passes the threshold of quantum detectors like the eye's photoreceptors, and can not be detected. One can notice that some species (like snakes) have localized special cells on their skin acting as detectors of IR-temporal variations. They are called "thermal eyes" and are used to detect moving preys.

1.1.1.3. *Indexes of Refraction*

When passing from one transparent medium to another, a ray of light is deviated: this is called "refraction". If i_1 is the angle of incidence and i_2 is the angle of refraction, the following relation holds: $n_1 \sin(i_1) = n_2 \sin(i_2)$, n_m being the refractive index of the medium m. It is related to the velocity of propagation of light in this medium "v", and to the velocity of propagation of light in vacuum "c":

$$n_m = \frac{c}{v} \tag{1.2}$$

The velocity of light depends on the medium

In vacuum, the velocity of light is independent of the frequency (or of wavelength). However, it varies according to the propagation medium. Table 1.1 shows some examples of the velocity of light in various media for a given wavelength $\lambda = 589\,\text{nm}$ (in vacuum).[2]

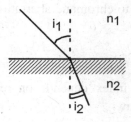

Figure 1.5. When an incident beam of light crosses the surface between two media, it is refracted according to a new angle, which depends on the ratio of the indexes.

---

[2]Wavelength is more useful as a unit ($\sim$500 nm) than frequency ($\sim 10^{15}$ Hz), because of its range of numerical values. But the velocity of light varies with media, an it would be misleading to speak of wavelength instead of frequency. For convenience, using the word "wavelength" without specification will always refer to the equivalent wavelength in vacuum.

Table 1.1. Velocity of light and indexes of refraction for various transparent media.

Medium	v (km/s)	n
vacuum	299 792	1.0000
air	299 705	1.0003
water (20°C)	224 850	1.3333
optical glass	197 000	1.5217
diamond	124 035	2.4169

Table 1.2. Variation of the index for a glass at three different wavelengths.

n	λ
1.521	486
1.515	589
1.513	656

The velocity of light depends on the wavelength

For a given medium, the velocity of light depends on the frequency (on the wavelength in vacuum), hence the refractive index. Table 1.2 gives the resulting indexes, at different wavelengths, for a given quality of glass. This means that at the interface between two media, the deflection of light rays depends on color, leading to chromatic aberrations in optical devices and in the eye.

The velocity of light depends on pressure and temperature

For gases, the refractive index depends on temperature and pressure, according to Gladstone's law:

$$n = 1 + K\rho \tag{1.3}$$

ρ being the mass of the gas per unit of volume.

1.1.2. *Sources of Light, Receptor Illumination*

In order to quantify the effects of light, it is important to define a set of measurement units. They are related to the *fluxes of energy* (J/s or W) emitted by sources, conveyed by light rays or impinging on receptive surfaces. We will consider the densities of fluxes per units of wavelength, surface, or solid angle (W/nm, W/m^2 or W/sr). Basically, we define the density of flux per unit area for a monochromatic light, as the element of the radiant flux dP (in all directions) that flows through an elementary area dS of a surface S, in W/m^2:

If the surface belongs to a receptor, the term used is "irradiance":

$$dP = E dS. \tag{1.4}$$

If it belongs to a source, the term used is: "excitance":

$$dP = M dS. \tag{1.5}$$

For a given spectrum of wavelengths, these quantities should be replaced by their densities per unit of wavelength integrated over the considered spectrum. For example:

$$M = \int_{\lambda} M_{\lambda}(\lambda) d\lambda. \tag{1.6}$$

If we consider the effects of visible light on the eye, these quantities will be weighted by the visibility function $V_{\lambda}(\lambda)$; and as a result, their name and units will change as stated in Table 1.3.

1.1.2.1. *Point Source*

Radiant and Luminous intensity

In the (theoretical) case of a point source located at O in space, we define the *radiant intensity* as the elementary flux of energy (power) dP radiated

Table 1.3. Radiometric and photometric quantities and their corresponding units. 1 W = 683 lm at 555 nm, 1 lm = 1 candela.steradian (cd.sr), cd will be defined later.

Radiometric units		Photometric units	
Radiant flux	W	Luminous flux	lumen (lm)
Irradiance	W/m^2	Illuminance	lm/m^2
Excitance	W/m^2		

in the direction $\vec{u}$ within an elementary solid angle $d\Omega$, in W/sr:

$$J(\vec{u}) = \frac{dP}{d\Omega}.$$

In case of visible light, it is called *luminous intensity*, measured in *candela* (cd).

The candela is the luminous intensity, in one determined direction, of a source emitting a monochromatic radiation with a wavelength of 555 nm, and a radiant intensity in that direction of $1/683$ W.sr^{-1}.

Irradiance

If we consider the effect of a point source on an elementary area dS' of a receptor surface S' at point O' with outer normal n', receiving an elementary power dP', the *irradiance* $E(u, n')$, in W/m^2, is defined from:

$$dP' = J(\vec{u})d\Omega = J(\vec{u})\frac{|\vec{u}.\vec{n}'|}{r^2}dS' = J(\vec{u})\frac{\cos(\theta')}{r^2}dS' = E(\vec{u}.\vec{n}')dS', \quad (1.7)$$

hence the irradiance of the receptor:

$$E(\vec{u}.\vec{n}') = J(\vec{u})\frac{\cos(\theta')}{r^2}, \quad \text{in W/m}^2, \quad (1.8)$$

r^2 being the distance between the source point O and the receptor point O'. In the case of visible light, the name is *illuminance* and the unit is the Lux (lm/m^2).

Remarks

The receptor surface receives less power if:

- it is far from the source,
- its area is small, or
- the angle of incidence θ' is large.

1.1.2.2. *Distant Compact Source*

If the distance to the source grows to infinite (Figure 1.6-b), the solid angle tends to 0 and if the received power is not negligible, the expression of

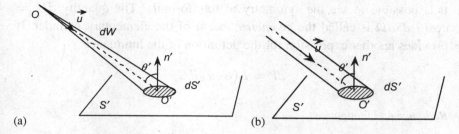

Figure 1.6. Diagrams for a) a point source and b) a distant source.

Table 1.4. Radiometric and photometric quantities.

Radiometric units		Photometric units	
Radiant intensity	W/sr	Luminous intensity	cd (=lm/sr)
Irradiance	W/m²	Illuminance	lux (=lm/m²)

irradiance (or illuminance) should be different and defined from:

$$dP' = I(\vec{u})\cos(\theta')dS' = E(\vec{u}, \vec{n}')dS', \tag{1.9}$$

thus:

$$E(\vec{u}, \vec{n}') = I(\vec{u})\cos(\theta'). \tag{1.10}$$

1.1.2.3. *Distributed Light Source*

If the light source is a surface I, the power impinging on an elementary surface of area dS' at O' will depend on the power radiated by all points O of the source, of elementary area dS. The above-defined radiant intensity should take a new form related to the element of area dS of the source, to its position and its normal direction. Then the expression of the elementary power received by the receptor surface is:

$$dP' = L(O, \vec{u})\cos(\theta)dSd\Omega = L(O, \vec{u})\cos(\theta)dS\frac{\cos(\theta')}{r^2}dS',$$

or

$$dP' = L(O, \vec{u})\frac{|\vec{u}, \vec{n}||\vec{u}, \vec{n}'|}{r^2}dSdS'. \tag{1.11}$$

It is possible to see the symmetry of this formula. The quantity $dU = \cos(\theta)dSd\Omega$ is called the *geometric extent* of the elementary cylinder. It provides another expression for the definition of the luminance:

$$dP = L(O,\vec{u})dU.$$

Radiance and luminance

The radiance from an elementary surface located at point O on the source is:

$$L(O,\vec{u})\cos(\theta)dS,$$

the term $L(O,\vec{u})$ is called *radiance*, it is measured in W/m^2/sr, or for visible light *luminance*, in lm/m^2/sr.

Remarks

Same remarks as for a point source hold, with, in addition:

- if the section of the elementary cylinder is kept constant while the angle θ increases, more source area is seen from O', and more power passes through the cylinder. This will be important for Lambertian reflection as will be seen later.
- the source is said to be *Lambertian* if $L(O,\vec{u})$ is independent of the direction $\vec{u}$.
- if the solid angle $d\Omega'$ under which the source is seen from the receptor is constant, then dP' is independent of the distance and proportional to the apparent surface $\cos(\theta')dS'$ of the receptor.

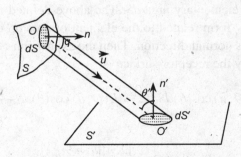

Figure 1.7.   Elementary cylinder between a source surface S and a receptor surface S'.

Table 1.5. Some photometric quantities for different usual light sources.

Illuminance		Luminance	
Dark night	10^{-4} lux	Minimum visible	10^{-5} cd/m^2
Star light	10^{-3} lux	Glowing worm	50 cd/m^2
Moon	10^{-1} lux	Flame	$15 \; 10^3$ cd/m^2
Office	300 lux	White paper in sun	$3 \; 10^4$ cd/m^2
Cloudy day	10^3 lux	Sun	$1.5 \; 10^9$ cd/m^2

As an example, Table 1.5 gives some values of illuminances and luminances for various sources of light.

1.1.3. *Reflection and Transparency*

1.1.3.1. *Specular Reflection*

Suppose that a ray of light hits the surface separating two transparent media (Figure 1.8) with indexes of refraction n_1 and n_2. The angle of incidence with respect to the normal n of the surface is i_1. The beam is generally split into a refracted beam (angle i_2) and a reflected beam (angle i_3), both situated in the plane of incidence.

The refraction angle is such that $n_1 \sin(i_1) = n_2 \sin(i_2)$, with the reflection angle being $i_3 = i_1$. The incident flux F_i of energy is also split into a reflected flux F_r, a transmitted flux F_t and an absorbed flux F_a. The ratios $R = F_r/F_i$ and $T = F_t/F_i$ are called *reflection* and *transmission factors* respectively. In case of no absorption, the relation $R + T = 1$ holds.

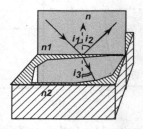

Figure 1.8. General case of reflection and transmission of light through a surface between two media of different indexes.

One must keep in mind that these relations are valid for *polished* surfaces such as glass, mirrors…, that is, for *specular* reflection. The case of glossy surfaces will be considered later.

Vitreous reflection

The reflection factor depends on the indexes n_1 and n_2 and also on the polarization state of the light (Figure 1.9). The direction of polarization can be decomposed into a component that is parallel to the plane of incidence and a component that is orthogonal. In fact there are two corresponding coefficients of reflection: respectively $R_\parallel$ and $R_\perp$. Therefore, the reflection on a vitreous surface changes the polarization of the incident light. As $R_\parallel < R_\perp$, the emergent ray tends to be more polarized perpendicularly to the plane of incidence.

- $R_\parallel$ presents a minimum (equal to 0) for a certain angle called "Brewsterian incidence angle", depending on the refractive index. For glass, it is around 55°. This is a means to generate polarized light.
- at normal incidence (0°), $R_\parallel$ and $R_\perp$ are equal: $R_\parallel = R_\perp = R_0$. For glass $R_0 = \frac{(n_1 - n_2)^2}{(n_1 + n_2)^2} = 0.04$ and $R_0 = 0.02$ for water at 20°C.
- A mean reflection factor is defined as $R = (R_\parallel + R_\perp)/2$.

The general formulae are:

$$R_\parallel = \frac{\tan^2(i_1 - i_2)}{\tan^2(i_1 + i_2)} \qquad T_\parallel = \frac{\sin(2i_1)\sin(2i_2)}{\sin^2(i_1 + i_2)\cos^2(i_1 - i_2)}$$

$$R_\perp = \frac{\sin^2(i_1 - i_2)}{\sin^2(i_1 + i_2)} \qquad T_\perp = \frac{\sin(2i_1)\sin(2i_2)}{\sin^2(i_1 + i_2)} \cdot g$$

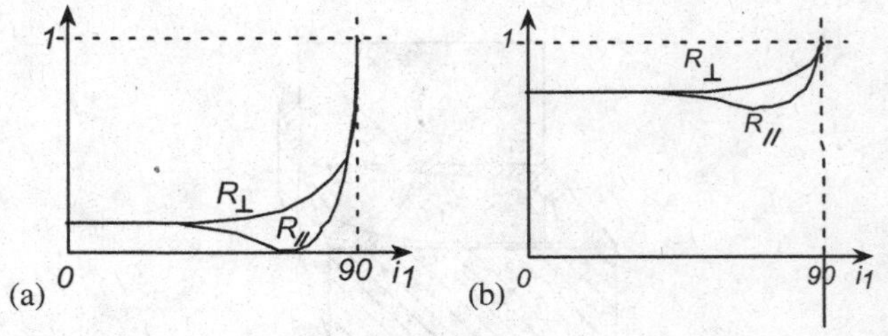

Figure 1.9. Vitreous (a) and metallic (b) reflection coefficients as functions of the angle of incidence. Note the minimum of the $R_\parallel$ coefficient for some incidence.

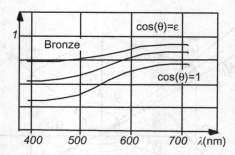

Figure 1.10. Reflection coefficient for bronze as a function of the wavelength and incidence angle.

Metallic reflection

The mean reflection factor of polished metals has a high value, even in case of normal incidence (Figure 1.9). It varies with wavelength. For example, silver has a reflection factor of 0.92 in visible light, 0.95 in IR and 0.1 in UV. Figure 1.10 gives the reflection factor for bronze versus the wavelength, at various incidence angles.

1.1.3.2. *Diffuse Reflection*

Diffuse reflection occurs for objects that reflect light in all directions, due to their surface roughness. The surface of an object can be considered both as a receptor with respect to incident light, and as a distributed source with respect to the reflected light (Figure 1.11).

As a receptor, the received power dP per unit of area dS from direction $\vec{u}_i$ in a solid angle $d\Omega_i$ is $\frac{dP}{dS} = L_i(\vec{u}_i)\cos(\theta_i)d\Omega_i$. The surface of an object is mainly characterized by what is called its "bi-directional reflectance distribution function" (BRDF): $\rho(\lambda, \vec{u}_i, \vec{u}_r)$, depending on all incident directions $\vec{u}_i$ and on the viewing direction $\vec{u}_r$. It also depends on the wavelength, hence it is responsible for the intrinsic color of objects.

As a source, the surface is characterized by its reflected luminance, that is, by the integration in all directions of incident light and its BRDF. Its luminance from a point O is:

$$L_r(\lambda, \vec{u}_r) = \int_{\Omega_i} L_i(\lambda, \vec{u}_i)\rho(\lambda, \vec{u}_i, \vec{u}_r)\cos(\theta_i)d\Omega_i. \qquad (1.12)$$

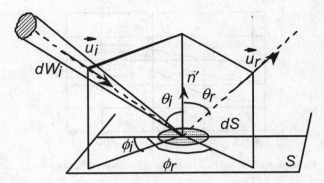

Figure 1.11.   Geometry for the general case of reflection. Light received from one direction is scattered in all directions according to the bi-directional reflection distribution function.

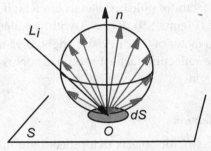

Figure 1.12.   For a Lambertian surface, the reflected luminance is proportional to the cosine of the reflection angle.

For Lambertian surfaces (Figure 1.12), the BRDF takes the simple form: $\rho(\lambda, \vec{u}_i, \vec{u}_r) = \rho_0(\lambda)$ where $\rho_0(\lambda)$ is called the "*albedo*" of the surface. Therefore, the power radiated by an elementary Lambertian surface dS is:

$$dP = L_r(\lambda)\cos(\theta_r)dS = L_r(\lambda)dS_a.$$

Notice that the formula (1.12) is a general one. For *specular* surfaces, the BRDF takes the following form:

$$\rho(\lambda, \vec{u}_i, \vec{u}_r) = R(\lambda, \theta_i)\delta(\theta_i - \theta_r)\delta(\phi_i - \phi_r + \pi).$$

Then, the reflected luminance is:

$$L_r(\lambda, \theta_r, \phi_r) = R(\lambda, \theta_i)L_i(\lambda, \theta_r, \phi_r - \pi),$$

Table 1.6. Reflection coefficient (albedo) for various usual materials.

Material	Albedo
Black velvet	0.004
White paper	0.75
Snow	0.95

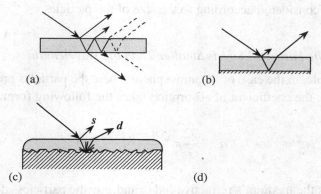

Figure 1.13. Some objects and their behavior with light. (a) a piece of glass, (b) a mirror, (c) a varnished glossy surface d) metallic cylinders on a table.

which shows that the reflected beam is in the plane of incidence, with $\theta_r = \theta_i$.

1.1.3.3. *Examples of Objects and Surfaces*

Table 1.6 gives the reflection factor (albedo) for various usual materials.

Figure 1.13 gives some examples of what kinds of reflection may happen to light according to different materials or objects such as glass, mirrors, furniture or painted cylinders.

1.1.4. *Atmospheric Diffusion*

Particles in suspension in a turgid medium reflect the rays of light in every direction (Figure 1.14). The transmitted flux depends on the density of the particles and on their size with respect to the wavelength of the light. It can be expressed by a general formula as a function of the thickness r of the

Figure 1.14.   Diffusion in a gas: Incident and transmitted fluxes.

medium: $\Phi_t = \Phi_i e^{-ar}$, a being the coefficient of absorption. Two cases should be considered according to the size of the particles.

1.1.4.1. *The Particles' Size is Smaller Than the Wavelength*

For example, in the case of the atmosphere where the particles are gaseous molecules, the coefficient of absorption takes the following form:

$$a = \frac{32}{3}\pi^3 \frac{(n-1)}{\rho} \frac{1}{\lambda^4}, \tag{1.13}$$

where n is the medium's refractive index and ρ is the particles' density.

The fact that it is a function of the inverse fourth power of λ is called the *law of Rayleigh*. This means that short wavelengths (when a is large) have a low transmission factor, and hence that they are strongly diffused. This explains why the sky is blue when observed in a direction different from that of the sun and why the sunset light is red. This means also that we are not completely protected against UV radiation even if we stay in the shade! Figure 1.15 illustrates the spectra of the transmitted and diffused sunlight. As an example, according to the thickness of the atmosphere ($r \sim$ 8000 m) and for a 400 nm wavelength, the transmitted flux is 73% of the incident one.

Moreover, for a particle situated in O (Figure 1.16), if the angle between the direction OM of observation and the direction Oz of propagation of the incident flux is θ, the diffused ray tends to be polarized: the component of the electric field orthogonal to the plane MOz is unchanged, but the component in the same plane is multiplied by $\cos(\theta)$. The diffused light is completely polarized in an observation direction orthogonal to the direction of the incident flux.

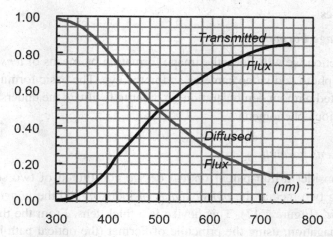

Figure 1.15. Transmitted light flux in the incident direction and diffused flux in all directions versus wavelength.

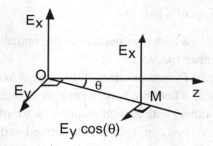

Figure 1.16. Polarization of light according to the direction of diffusion.

1.1.4.2. *The Particles' Size is Larger Than Wavelength*

For example, in the case of a cloud where the particles are droplets considered as little spheres of water of radius $R \sim 5\mu$, the absorption coefficient does not depend on the wavelength and takes the simplified form: $a = 2\pi\rho R^2$, where ρ is the density of the water droplets.

Thus, for a cloud with 300 droplets per cm^3, with a thickness of 100 m, the transmission ratio is 1/100. This means that almost all the incident light is diffused. As the diffusion does not depend on wavelength, the clouds look white on the blue background of the sky.

1.2. Optics

1.2.1. *Image Formation*

In this section we will study the image formation by means of *lenses*, as it occurs in photo and video cameras and in the eye. The basic formulae will be presented without demonstration, in order to facilitate the understanding of the various phenomena.

1.2.1.1. *Lens Formula*

Let us consider a transparent optical device constituted of two surfaces, separating two media: the *object space*, of index n, and the *image space*, of index n' (Figure 1.17). This device is a thick lens. From the theory of light propagation, using the principle of Fermat (the optical path is either maximal or minimal) and the approximation of Gauss (light rays are slightly inclined with respect to the optical axis), a general formula can be obtained as a relation between the distances of *cardinal points*.

The cardinal points of a lens are:

- *Focal Points* F, F': where rays parallel to the optical axis in one space converge in the other space.
- *Principal Points* P, P': intersection of the optical axis by the *principal planes*, the surfaces where the wave fronts are conjugate planes (one is the image of the other) with a magnification factor of 1.
- *Nodal Points* N, N': two conjugate points defined so that, to any incident ray passing through N, will correspond an emergent ray parallel to the incident one and passing through N'.

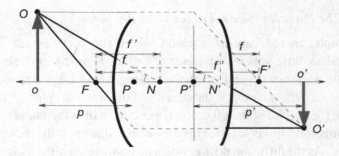

Figure 1.17.   Diagram of a thick lens.

The characteristic distances are:

- *Focal lengths f, f'*: $f = FP$ and $f' = F'P'$, they are defined by the geometry of the surfaces and by the indexes of refraction. Their ratio is obtained by: $f/f' = n/n'$, that is, the ratio between the indexes of object and image spaces. In camera optics, as the two spaces are made of the same medium (air), $f = f'$. This is not the case for the eye where the image space is a biological liquid, then the ratio is rather: $f/f' = 0.75$.
- *Distances between nodal, focal and principal points.* We have the following relations: $NP = N'P'$, $N'F' = f$ and $FN = f'$.
- *Object* and *Image distances*: they are defined with respect to the object and image principal planes, the *object distance* is $p = oP$, the *Image distance* is $p' = o'P'$.

The relation between image and object distances is derived from geometrical considerations:

$$\frac{f}{p} + \frac{f'}{p'} = 1, \quad \text{with} \quad \frac{f}{f'} = \frac{n}{n'}.$$

1.2.1.2. Thin Lens

For usual cameras (even with thick lenses), we have $n = n'$, which implies that the two focal lengths are equal ($f' = f$), then $N = P$ and $N' = P'$. In addition, for *thin lenses* we can neglect the thickness of the lens, which leads to $P = P'$ and $N = N'$.

Given these particular cases, the formulae and the construction of the image O' of an object O become much simpler. The basic formula is:

$$\frac{1}{p} + \frac{1}{p'} = \frac{1}{f}. \tag{1.14}$$

If the lens is made of a material of index n, its two faces being spheres of radii R_1 *and* R_2, its focal length is given by the lens makers' equation:

$$\frac{1}{f} = (n - 1) \left(\frac{1}{R_1} + \frac{1}{R_2} \right).$$

The magnification factor is:

$$\gamma = \frac{p'}{p}.$$

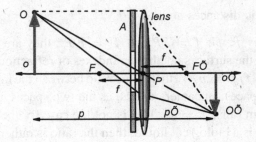

Figure 1.18.   The geometry for a thin lens.

Note that by modifying the radii of curvature it is possible to adjust the focal length. This is naturally made in the eye, by means of muscles that stretch the elastic crystalline lens (though it is not really a thin lens). The diagram of a thin lens is given at Figure 1.18.

The aperture A of the lens (e.g. pupil for the eye) can be modeled as being an occluder within the principal plane. Practically, in cameras (diaphragm) it is situated near one face of the lens. For the eye, the occluder (iris) is close to the lens, on the object side.

1.2.2. *Image on a Receptor Surface*

1.2.2.1. *Focusing*

From the formulae, it can be seen that, for a given focal length, the position p' where the image is formed depends on the position p of the object. If we have a mosaic of receptors to capture the image, the position of this mosaic should move to keep the objects in focus. In most optical systems (cameras or eyes), the surface where the image is formed is not movable. The solution for cameras is to move the lens by an amount of Δx such that $p' \to p' + \Delta x$ and $p \to p - \Delta x$, keeping the image-object distance constant. The solution for the eye is to stretch the crystalline lens in order to modify the focal length, thus keeping the nodal distance (principal plane to image) constant, and the eye's dimensions as well!

1.2.2.2. *Depth of Field*

If the lens (Figure 1.19) has a pupil of diameter D, it is seen from the image point o' under the angle α' such that: $\tan(\alpha') = D/p'$. If the diameter of

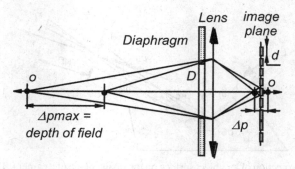

Figure 1.19. Depth of field.

a photoreceptor is d, the allowed defocus is such that: $\Delta p'/p' < d/D$. As from the thin lens formula we derive: $\Delta p' = p'^2/p^2 \Delta p$, we see that the maximum variation of object distance without perceptible defocus is:

$$\Delta p_{max} = 2\frac{p^2}{p'^2}\frac{d}{D}. \tag{1.15}$$

This means that Δp_{max} is the range allowed for p without perceptible defocus. The term Δp_{max} is called the *depth of field*. For a given receptor diameter, it increases with a greater distance to the object and with the narrowing of aperture. In the case of the eye, objects are more acute in full light than in dim light, this is due to the pupilar reflex that diminishes the diameter of the pupil when the light intensity is strong, thus augmenting the depth of field.

1.2.3. *Photometry and Lenses*

1.2.3.1. *Image of a Lambertian Surface*

The aperture dA of the lens of Figure 1.20 receives a flux:

$$dP = L(O, \vec{u}_c)dU = L(O, \vec{u}_c)\cos(\theta_c)dAd\Omega$$

The same flux, multiplied by the transmittance of the lens T_l is received by the pixel x_i which views the aperture dA under $d\Omega_i$:

$$dP_i = T_l L(O, \vec{u}_c)\cos(\theta_c)dAd\Omega_i,$$

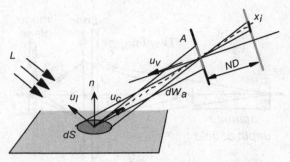

Figure 1.20.   Projection of an object surface on the image plane of a camera. The luminance of the pixel area will depend on the aperture and on the viewing angle.

expressing $d\Omega_i$ as a function of the pixel surface dS_i viewed from the nodal point:

$$dP_i = T_l L(O, \vec{u}_c) \cos(\theta_c) dA \frac{\cos(\theta_c) dS_i}{r^2},$$

and introducing $r = ND/\cos(\theta_c)$, ND being the nodal distance:

$$dP_i = T_l L(O, \vec{u}_c) dA \frac{\cos^4(\theta_c) dS_i}{ND^2}. \qquad (1.16)$$

Note that the image irradiance:

- does not depend on the distance to the surface
- falls off like $\cos^4(\theta_c)$ in the corners of the image, θ_c being the angle between the viewing direction and the camera's axis. Therefore, for wide-angle images, there is a significant roll-off in the image intensity towards edges and corners.[3]

1.2.3.2. *Invariance of Optical Extent*

The optical extent is the geometrical extent multiplied by the squared refractive index: $n^2 U$.

According to the theorem of Clausius, "For any optical system, be it stigmatic or not, the optical extent is invariant from object space to

---

[3] For the eye, due to the sphericity of the retinal surface, the fall-off is like $\cos(\theta_c)$, which is much more efficient.

image space":

$$n^2 U = n'^2 U'.$$

This means that for a Lambertian surface of luminance L radiating a power P, the luminance L' of the image is obtained as follows. The total power of the image is $P' = L'U' = T_l P = T_l L U$, T_l being the transmittance of the lens. Then, the luminance of the image is:

$$L' = T_l \frac{n'^2}{n^2} L.$$

For the eye, we have: $L' = 1.78\, T_l L$, which provides more efficiency than for cameras where $n'/n = 1$.

1.2.4. *Physical Limitations to Resolution*

1.2.4.1. *Defocus Blur*

An improperly focused lens causes a point-spread function (PSF) around the axis of projection. This PSF can be modeled at first order as a disk of constant level (geometrical considerations can be used to get a rough estimate of the size of the disk). If $g_d(x, y)$ is the PSF, then the resulting image $i_d(x, y)$ is the convolution of the ideal image $i_i(x, y)$ by the kernel $g_d(x, y)$, that is: $i_d(x, y) = i_i(x, y) * g_d(x, y)$.

1.2.4.2. *Motion Blur*

If there is some temporal averaging, the image of a moving point forms a streak in the image, giving a blurry image (Figure 1.21).

1.2.4.3. *Diffraction*

Even a properly focused point is not imaged to a point. Rather, there is a point-spread function (PSF). For diffraction alone, this PSF can be modeled

Figure 1.21. Blur effects on an image. Successively: Original, Gaussian, defocus and motion.

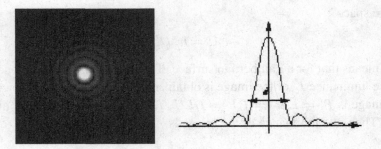

Figure 1.22.   Diffraction: Airy disk and its section.

using the "Airy disk" (Figure 1.22), which has a diameter of:

$$d = \frac{1.22\lambda}{n'}\frac{\overline{ND}}{D}.$$

For example, for human eyes (see Wyszecki and Stiles, *Color Science*, 1982):

- the index of refraction within the eye is $n' = 1.33$,
- the nodal distance is $\overline{ND} \approx 17$ mm,
- we generally use $\lambda = 500$ nm,
- the pupil diameter is $d \approx 2$ mm (adapted to bright conditions).

Therefore the diameter of the Airy disk is $d \sim 4.4\,\mu$.

This compares to the distance between foveal cones, which is $\sim 2.5\,\mu$. It shows that human vision operates near the diffraction limit.

By the way, a 2-pixel spacing in the human eye corresponds to having a 128×128 pixel resolution of the image of your thumbnail at arm's length. If we compare this image accuracy to the typical sizes of images used by machine vision systems, we notice that it is at least twice the accuracy of a 512×512 pixels image seen on the computer screen.

1.2.4.4. *Photon Noise*

The average photon flux (spectral density) at the image (in units of photons per sec, per unit wavelength) is equal to the irradiant flux divided by the energy of one photon: $\varphi = P/h\nu$. The photon arrivals follow a Poisson

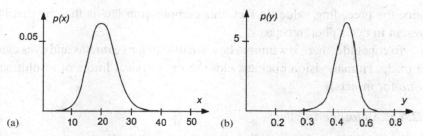

Figure 1.23. (a) probability density function for a Poisson law of mean $\mu_x = 20$, (b) probability density function after the compression. Notice the narrower profile.

statistics (the variance is equal to the mean photon catch):

$$P[k|\varphi T] = \frac{(\varphi T)^k}{k!}e^{-\varphi T},$$

with mean $\mu = \varphi T$ and standard deviation $\sigma = \sqrt{\varphi T}$.

There is a significant photon noise component with a standard deviation linked to the signal. In the example of Figure 1.23, the signal-to-noise ratio (SNR) is $\text{SNR}_x = 4.47$.

Here, the traditional assumptions about signal and noise relationships do not hold, making the estimation of noise effects more complicated:

- photon noise is not Gaussian,
- it is not independent of the signal,
- it is not additive.

If we apply a compression law on the amplitude of the signal, it is possible to improve the SNR. Let us apply the following transformation on the signal x:

$$y = \frac{x}{x + \mu_x} = f(x),\tag{1.17}$$

μ_x being obtained from some time or spatial average of the signal. Now, the probability density function of the new signal y is:

$$p(y) = (f^{-1}(y))'_y \frac{(\mu_x)^{f^{-1}(y)}}{(f^{-1}(y))}e^{-\mu_x},$$

the profile of which is given in Figure 1.23. From this formula, the new signal to noise ratio can be computed, its value is $\text{SNR}_y = 8.616$, that is, twice the preceding value. In fact, this compression law is the one that is present in eye's photoreceptors.

To conclude: there is a limit to how small standard cameras and eyes can be made. Human vision operates close to the physical limits of resolution (*ditto* for insects).

1.2.4.5. *Atmospheric Effect*

In fact, the diffusion process in the atmosphere is not the unique source of image perturbation. The turbulences always present in the air make the local refractive index vary. This causes a blurred image depending on the thickness of the traversed air layer. This effect is often modeled as the filtering of the image by a Gaussian kernel. Just think how the mountains seem nearer on a winter day when the air is dry and cold: there are few turbulences and little diffusion.1.3 From 3D World to 2D Images.

In the preceding sections, we have studied the transformation of light (spectral composition and intensity) with respect to the traversed media, the reflections of objects and the optical devices. Let us now study how the geometrical relations between points of the (2D) image surface are linked to the geometrical relations between points of objects in the (3D) volume of the world. For this purpose, we will analyze the fundamentals of perspective projection.

1.3. From 3D World to 2D Images

1.3.1. *The Pinhole Camera and Its Model*

In the pinhole camera, we choose a coordinates system (x, y, z) with its origin at the pinhole. The image $m(x, y, -d)$ of a point $M(x, y, z)$ is inverted and scaled by the ratio or magnifying factor d/z (Figure 1.24, Left), d being the nodal distance between the image plane and the pinhole.

This inversion may lead to confusions for further calculations, hence the non-inverting model for mathematical convenience (Figure 1.24-b). We choose a world coordinates system (X, Y, Z) with an object point $M(X, Y, Z)$. Then we place an image plane at a distance $-d$ from the origin.

1.3.2. *Basic Equations, Homogeneous Coordinates*

Let $M(X, Y, Z)$ be a 3D point: its image position $m(x, y)$ in a coordinate system linked to the image plane is given by the classical projective

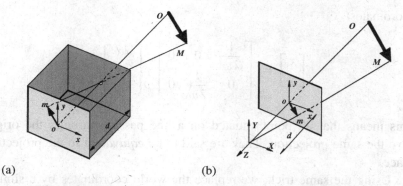

Figure 1.24. (a) the pinhole camera and its inverted image. (b) the model of conic projection with its non-inverted image.

transformation formulae:

$$x = \frac{f}{Z}X, \quad y = \frac{f}{Z}Y. \tag{1.18}$$

Here, f stands for d. It results from the confusion between nodal distance and focal distance, usual in computer science. We should remember that if we use optics, we should replace f by the nodal distance for correct numerical applications.

The problem is that these equations for coordinates in the Euclidean space of dimension 2 are non-linear and we can easily imagine the resulting complications in mathematical developments. For this reason, we prefer to use another system of coordinates called *homogeneous coordinates*. They are linked to a projective space of dimension $2 + 1$. Let us add a third dimension to the vector of the image plane: $(x, y) \rightarrow (x, y, w)$. When necessary, the original image coordinates can be obtained from the following matrix transform applied to (x, y, w):

$$\begin{bmatrix} x \\ y \end{bmatrix} \Leftarrow \begin{bmatrix} \dfrac{1}{w} & 0 & 0 \\ 0 & \dfrac{1}{w} & 0 \end{bmatrix} \cdot \begin{bmatrix} x \\ y \\ w \end{bmatrix}. \tag{1.19}$$

Note that from this, a homogeneous coordinate can be multiplied by any constant different from 0, without changing the result in the image space coordinates:

$$\begin{bmatrix} x \\ y \end{bmatrix} \Leftarrow \begin{bmatrix} \dfrac{1}{\lambda w} & 0 & 0 \\ 0 & \dfrac{1}{\lambda w} & 0 \end{bmatrix} \cdot \begin{bmatrix} \lambda x \\ \lambda y \\ \lambda w \end{bmatrix}.$$

This means that all point situated on a line passing through the origin have the same projection. They are said to be *equivalent* in the projective space.

Using the same trick, we replace the world coordinates by a similar homogeneous coordinates system $(X, Y, Z) \rightarrow (X, Y, Z, W)$ and thus, the projection formulae relating the world projective space to the camera projective space take a matrix form $\mathbf{x} = \mathbf{A} \cdot \mathbf{X}$, expanding to:

$$\begin{bmatrix} x \\ y \\ w \end{bmatrix} = \begin{bmatrix} f & 0 & 0 & 0 \\ 0 & f & 0 & 0 \\ 0 & 0 & 1 & 0 \end{bmatrix} \cdot \begin{bmatrix} X \\ Y \\ Z \\ W \end{bmatrix}.$$

Now suppose that the camera is rotated around its axes by angles $(\theta_x, \theta_y, \theta_z)$, resulting in a rotation matrix R_c, and translated by a vector $t_c^T = [t_x, t_y, t_z]$. According to these modifications, we can re-write the preceding equation, using a 3×3 matrix $\mathbf{A}$ for the camera and a 3×4 matrix for the rotation and translation:

$$\begin{bmatrix} x \\ y \\ w \end{bmatrix} = \begin{bmatrix} f & 0 & 0 \\ 0 & f & 0 \\ 0 & 0 & 1 \end{bmatrix} \cdot \begin{bmatrix} r_{11} & r_{12} & r_{13} & t_x \\ r_{21} & r_{22} & r_{23} & t_y \\ r_{31} & r_{32} & r_{33} & t_z \end{bmatrix} \cdot \begin{bmatrix} X \\ Y \\ Z \\ W \end{bmatrix} = \mathbf{A} \cdot \mathbf{RT} \cdot \begin{bmatrix} X \\ Y \\ Z \\ W \end{bmatrix}, \tag{1.20}$$

and, replacing W by 1, we have:

$$\begin{bmatrix} x \\ y \\ w \end{bmatrix} = \begin{bmatrix} f & 0 & 0 \\ 0 & f & 0 \\ 0 & 0 & 1 \end{bmatrix} \cdot \begin{bmatrix} R_1(X, Y, Z) + t_x \\ R_2(X, Y, Z) + t_y \\ R_3(X, Y, Z) + t_z \end{bmatrix}, \tag{1.21}$$

then, returning to the non homogeneous coordinates in the image space:

$$\begin{cases} x = \dfrac{f}{R_3(X,Y,Z) + t_z}[R_1(X,Y,Z) + t_x] \\[3mm] y = \dfrac{f}{R_3(X,Y,Z) + t_z}[R_2(X,Y,Z) + t_y] \end{cases} \quad (1.22)$$

Note that:

- the matrix **A** holds for the *internal parameters* of the camera. In particular, if there are different magnification factors on x and y axes, some skewing and a displacement of the origin in the camera plane (x_0, y_0), the new form of **A** is:

$$\mathbf{A} = \begin{bmatrix} fa_{11} & fa_{12} & x_0 \\ 0 & fa_{22} & y_0 \\ 0 & 0 & 1 \end{bmatrix}.$$

- the matrix **RT** holds for the *external parameters* of the camera.

1.3.3. *Linear Algebra in Homogeneous Coordinates*

1.3.3.1. *One Point on a Line*

In the image plane, a point $\mathbf{x}' = [x_1', x_2']$ on a line satisfies the following equation: $l_1 x_1' + l_2 x_2' + l_3 = 0$. Which relation does a corresponding point in the projective space satisfy? In the 3D space of homogeneous coordinates, a 3-vector represents the corresponding point: $\mathbf{x} = [x_1, x_2, x_3]$. Recall that the correspondence between the two points is:

$$[x_1', x_2'] = \left[\frac{x_1}{x_3}, \frac{x_2}{x_3}\right],$$

and, by substitution in the equation of the line $l_1 \frac{x_1}{x_3} + l_2 \frac{x_2}{x_3} + l_3 = 0$, we finally obtain:

$$l_1 x_1 + l_2 x_2 + l_3 x_3 = 0. \quad (1.23)$$

Hence a line in the image plane corresponds to the equation of a plane in the projective space. However we will speak of a *line* and define it by its parameters: $\mathbf{l} = [l_1, l_2, l_3]$. Any point belonging to the line satisfies the equation $l_1 x_1 + l_2 x_2 + l_3 x_3 = 0$, which can be written as a dot product: $\mathbf{l}^T \mathbf{x} = \mathbf{x}^T \mathbf{l} = 0$.

In the same manner, we will define the distance of a point from a line by $d = l_1 x_1 + l_2 x_2 + l_3 x_3$. This will correspond to the Euclidean distance in the image plane, only if we normalize the line parameters by $l_i \rightarrow l_i / \sqrt{l_1 + l_2}$ and the coordinates of the point by $x_i \rightarrow x_i / x_3$.

1.3.3.2. *Two Points Define a Line*

Let $\mathbf{l} = [l_1, l_2, l_3]$ be a line passing through two points $\mathbf{p} = [p_1, p_2, p_3]$ and $\mathbf{q} = [q_1, q_2, q_3]$. As we have seen, the line parameters must satisfy both equations: $l_1 p_1 + l_2 p_2 + l_3 p_3 = 0$ and $l_1 q_1 + l_2 q_2 + l_3 q_3 = 0$.

A compact way to write the solution of these equations is to use the cross product: $\mathbf{l} = \mathbf{p} \times \mathbf{q}$, that is:

$$\begin{bmatrix} l_1 \\ l_2 \\ l_3 \end{bmatrix} = \begin{bmatrix} p_2 q_3 - p_3 q_2 \\ p_3 q_1 - p_1 q_3 \\ p_1 q_2 - p_2 q_1 \end{bmatrix}, \text{ or in matrix form: } \mathbf{l} = [\mathbf{p}]_\times \mathbf{q},$$

with:

$$[\mathbf{p}]_\times = \begin{bmatrix} 0 & -p_3 & p_2 \\ p_3 & 0 & -p_3 \\ -p_2 & p_3 & 0 \end{bmatrix}. \tag{1.24}$$

Note that $[\mathbf{p}]_\times$ is a 3×3 matrix of rank 2.

1.3.3.3. *Two Lines Define a Point*

In the same idea, two lines $\mathbf{l} = [l_1, l_2, l_3]$ and $\mathbf{m} = [m_1, m_2, m_3]$ intersect at a point $\mathbf{x} = [x_1, x_2, x_3]$.

The solution is given by $\mathbf{x} = [\mathbf{l}]_\times \mathbf{m}$,

$$\text{that is: } \mathbf{x} = [\mathbf{l}]_\times \mathbf{m} \Rightarrow \begin{bmatrix} x_1 \\ x_2 \\ x_3 \end{bmatrix} = \begin{bmatrix} l_2 m_3 - l_3 m_2 \\ l_3 m_1 - l_1 m_3 \\ l_1 m_2 - l_2 m_1 \end{bmatrix}.$$

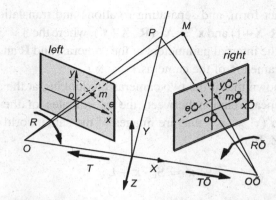

Figure 1.25. Left: the pinhole camera and its inverted image. Right: the model of conic projection with its non-inverted image.

1.3.4. *Stereovision*

Two cameras (left and right) with parallel vertical axes have been respectively rotated by $\mathbf{R}$ and $\mathbf{R}'$, and translated by $\mathbf{T}$ and $\mathbf{T}'$ (Figure 1.25). Their foci O and O' are at the same height, they are fixating the same point P. The left and right images of P are respectively o and o'.

A point $\mathbf{M} = [X, Y, Z]$ in 3D space will be projected in each image plane respectively in m and m'. In the two camera systems of coordinates, the components of these two points are different: $m = [x, y]$ and $m' = [x', y']$.

1.3.4.1. *Basic Formulae*

Taking the origin of world coordinates at the middle point between O and O', we can apply to each camera the general formulae:

$$
\begin{bmatrix} x \\ y \end{bmatrix} \Leftarrow
\begin{bmatrix} \dfrac{1}{w} & 0 & 0 \\ 0 & \dfrac{1}{w} & 0 \end{bmatrix} \cdot
\begin{bmatrix} x \\ y \\ w \end{bmatrix},
$$

with:

$$
\begin{bmatrix} x \\ y \\ w \end{bmatrix} =
\begin{bmatrix} f & 0 & 0 \\ 0 & f & 0 \\ 0 & 0 & 1 \end{bmatrix} \cdot
\begin{bmatrix} r_{11} & r_{12} & r_{13} & t_x \\ r_{21} & r_{22} & r_{23} & t_y \\ r_{21} & r_{22} & r_{23} & t_z \end{bmatrix} \cdot
\begin{bmatrix} X \\ Y \\ Z \\ W \end{bmatrix},
$$

where W can be replaced by 1.

In a compact form, and separating rotations and translations, we can write: $\mathbf{x} = \mathbf{A} \cdot (\mathbf{R} \cdot \mathbf{X} + \mathbf{t})$ and $\mathbf{x}' = \mathbf{A}' \cdot (\mathbf{R}' \cdot \mathbf{X} + \mathbf{t}')$, where the 3×3 matrices $\mathbf{A}$ and $\mathbf{A}'$ include the internal parameters of the cameras, and $\mathbf{R}$ and $\mathbf{R}'$ include the external parameters of the cameras.

It can be shown that if the two cameras are looking at the same scene, there exist a linear relation between the coordinates of the two points $\mathbf{x}(x, y, w)$ and $\mathbf{x}'(x', y', w')$ that are images of the same world coordinates point $\mathbf{X}(X, Y, Z, W)$:

$$\mathbf{x}' = \mathbf{S}_c \cdot \mathbf{x} + \mathbf{t}_c. \tag{1.25}$$

1.3.4.2. *Essential and Fundamental Matrices*

Working in the projective space, we can consider that the (normalized) homogeneous coordinates of a point $\mathbf{m}' = [m_1', m_2', 1]$ in one camera are related to those of point $\mathbf{m} = [m_1, m_2, 1]$ in the second one by a rotation $\mathbf{R}_c$ and a translation $\mathbf{t}_c$, that is: $\mathbf{m}' = \mathbf{R}_c \mathbf{m} + \mathbf{t}_c$.

In order to express that the four points O, O', $\mathbf{m}$ and $\mathbf{m}'$ are coplanar, we express that the cross product of $\mathbf{t}_c$ by the rotated version of $\mathbf{m}$ is orthogonal to $\mathbf{m}'$. Taking the cross product by $\mathbf{t}_c$, we have: $[\mathbf{t}_c]_\times \mathbf{m}' = [\mathbf{t}_c]_\times \mathbf{R}_c \mathbf{m}$, and taking the dot product by $\mathbf{m}'$, we finally obtain: $\mathbf{m}'^T [\mathbf{t}_c]_\times \mathbf{m}' = \mathbf{m}'^T [\mathbf{t}_c]_\times \mathbf{R}_c \mathbf{m} = 0$, which is called the *epipolar constraint*. This results in two linear equations. The following matrix:

$$\mathbf{E} = [\mathbf{t}_c]_\times \mathbf{R}$$

is called the *essential matrix*. If the cameras are not calibrated, we should introduce the calibration matrices $\mathbf{C}$ and $\mathbf{C}'$ of the cameras (such as $\mathbf{m} \rightarrow \mathbf{C}^{-1} \mathbf{m}$ and $\mathbf{m}' \rightarrow \mathbf{C}'^{-1} \mathbf{m}'$). This will change $\mathbf{E}$ into the new form $\mathbf{F}$ called the *fundamental matrix*:

$$\mathbf{F} = \mathbf{C}'^{-T} \mathbf{E} \mathbf{C}^{-1}.$$

The 3×3 essential matrix $\mathbf{E}$ has 5 degrees of freedom: 3 for rotation and 2 for translation (due to depth ambiguity, only 2 translation parameters are possible). There are two additive constraints: its determinant is 0 and its two non-zero singular values are equal. But because of the normalization, all its elements can be scaled by the last one e_{33}, which leaves only 8 degrees of freedom. The 3×3 fundamental matrix $\mathbf{F}$ has 7 degrees of freedom and its rank is also 2.

Essential and fundamental matrices are used in problems for finding corresponding points between two images in stereovision or between two successive images in motion estimation.

1.3.4.3. *Epipolar Lines*

Recall the general equation of a line $l_1 x_1 + l_2 x_2 + l_3 x_3 = 0$ and compare it to the epipolar constraint $\mathbf{m}'^T [\mathbf{t_c}]_\times \mathbf{R} \mathbf{m} = 0$ or $\mathbf{m}'^T \mathbf{F} \mathbf{m} = 0$. This means that, in the second image, all the possible points $\mathbf{m}'$ corresponding to $\mathbf{m}$ in the first one lie on a line, the parameters of which are defined by: $\mathbf{e}' = \mathbf{F}\mathbf{m}$. Such a line is called the *epipolar line* corresponding to $\mathbf{m}$. Similarly, there is an epipolar line in the first image, corresponding to the point $\mathbf{m}'$. Its parameters are: $\mathbf{e} = \mathbf{F}^T \mathbf{m}'$. The intersections of the epipolar lines with the line joining O and O' are called the *epipoles*.

1.3.5. *Perspective Transformation of Surfaces*

1.3.5.1. *Transformation of Coordinates*

With respect to the camera, the following coordinates will be used (Figure 1.26). World coordinates: $(x, y, z)_w$, image coordinates: $(x, y)_i$, and object surface coordinates: $(x, y)_s$. As usual, the image lies at a nodal distance f from the origin (the center of the image is at $-f$).

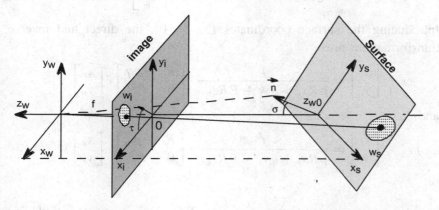

Figure 1.26. Model of the conic projection of a tilted and slanted surface.

The viewed surface is represented by its tangent plane (the equation of which is $z_w = px_w + qy_w + z_{w0}$) that intersects the optical axis z_w at point z_{w0}. The normal to the surface $\vec{n} = [p, q, 1]^T$ makes an angle σ with the direction of sight (slant), and an angle τ with the horizontal axis of the camera (tilt). We have: $\cos(\sigma) = 1/\sqrt{1 + p^2 + q^2}$ and $\tan(\tau) = q/p$, or conversely: $p = -\tan(\sigma)\cos(\tau)$ and $q = -\tan(\sigma)\sin(\tau)$. In the sequel, we will use the following variables: $P = \sqrt{1 + p^2}$ and $R = \sqrt{1 + p^2 + q^2}$.

We define coordinates linked to the surface plane, where x_s is the orthographic projection of x_w on the surface, z_s the normal to the surface and y_s is chosen in order to have a rectangle trihedron:

$$\begin{bmatrix} x_s \\ y_s \\ z_s \end{bmatrix} = \begin{bmatrix} (1 & 0 & p)/P \\ (-pq & P^2 & q)/PR \\ (-p & -q & 1)/R \end{bmatrix} \cdot \begin{bmatrix} x_w \\ y_w \\ z_w \end{bmatrix}.$$

Any rotation around z_s could also be convenient. The transformation from the surface coordinates to the world coordinates is:

$$\begin{bmatrix} x_w \\ y_w \\ z_w \end{bmatrix} = \begin{bmatrix} 1/P & -pq/PR & -p/R \\ 0 & P/R & -q/R \\ p/P & q/PR & 1/R \end{bmatrix} \cdot \begin{bmatrix} x_s \\ y_s \\ z_s \end{bmatrix} + \begin{bmatrix} 0 \\ 0 \\ z_{w0} \end{bmatrix}.$$

We remember that the perspective projection on the image plane is:

$$\begin{bmatrix} x_i \\ y_i \end{bmatrix} = \frac{f}{z_w} \cdot \begin{bmatrix} 1 & 0 & 0 \\ 0 & 1 & 0 \end{bmatrix} \cdot \begin{bmatrix} x_w \\ y_w \\ z_w \end{bmatrix}.$$

Introducing the surface coordinates ($z_s = 0$), the direct and inverse transformations are:

$$\begin{bmatrix} x_i \\ y_i \end{bmatrix} = \frac{f}{pRx_s + qy_s + PRz_{w0}} \cdot \begin{bmatrix} R & -pq \\ 0 & P^2 \end{bmatrix} \cdot \begin{bmatrix} x_s \\ y_s \end{bmatrix},$$

and

$$\begin{bmatrix} x_i \\ y_i \end{bmatrix} = \frac{f}{pRx_s + qy_s + PRz_{w0}} \cdot \begin{bmatrix} R & -pq \\ 0 & P^2 \end{bmatrix} \cdot \begin{bmatrix} x_s \\ y_s \end{bmatrix},$$

or:

$$\mathbf{x}_i = \frac{1}{a_s(x_s, y_s)} \mathbf{T} \cdot \mathbf{x}_s, \tag{1.26}$$

and:

$$\mathbf{x}_i = \frac{1}{a_s(x_s, y_s)} \mathbf{T} \cdot \mathbf{x}_s, \tag{1.27}$$

with the relation: $a_s(x_s, y_s) = RP^2/a_i(x_i, y_i)$ and:

$$a_i = \frac{P(px_i + qy_i - f)}{-z_{w0}} = \frac{Pf(px_w + qy_w - z_w)}{-z_{w0}z_w} = \frac{Pf}{z_w}.$$

For visible points on a given plane, the sign of a_i does not change and its variation is monotone:

$$\nabla_{a_i} = -\frac{P}{z_{w0}} \begin{bmatrix} p \\ q \end{bmatrix}, \quad \text{or} \quad \frac{\nabla_{a_i}}{a_i} = \frac{\begin{bmatrix} p & q \end{bmatrix}^t}{px_i + qy_i - f} \tag{1.28}$$

What these transformation formulae tell us:

- $\mathbf{T}$ is an affine transformation of the form $\mathbf{T} = \mathbf{R}_1.\mathbf{D}.\mathbf{R}_2$, with 3 degrees of freedom.
- a_i and a_s are zoom factors depending on the distance of view, hence on the position in the image.
- the gradients of a_i is in the direction of the normal's projection (p, q).

What is the use of these formulae?

They are useful to understand how a texture "painted" on the surface maps to the image plane. For example, if the viewed surface is a homogeneous texture, the variation of the zoom factor indicates the direction of the normal to the surface (see shape from texture). If the surface is composed of parallel lines, the direction of the normal will be estimated from the matrix $\mathbf{T}$.

We should remember that:

- The surface coordinates are defined up to some rotation around z_s.
- Ocular movements are equivalent to a rotation of coordinates $(x, y, z)_w$ around the axis x_w for vertical movements and around the axis y_w for horizontal movements.

1.3.5.2. *Image of the Local Properties of the Viewed Surface*

Let us consider a given object in the scene. The equation of its visible surface is: $z_w = f(x_w, y_w)$ or $f(x_w, y_w) - z_w = 0$ in world coordinates. Let $[dx, dy, dz]_w$ be a small variation around a given position, then we have:

$$\left(\frac{\partial f}{\partial x}dx + \frac{\partial f}{\partial y}dy - dz\right)_w = 0, \quad \text{that is} \quad (f'_x dx + f'_y dy - dz)_w = 0$$

or:

$$[f'_x, f'_y, -1]_w^T \cdot [dx, dy, dz]_w = 0. \tag{1.29}$$

This is the equation of the plane tangent to the surface at $[x, y, z]_w$, the normal of which is: $\mathbf{n} = [f'_x, f'_y, -1]^t = [p, q, -1]^t$. The radiance of the surface element is related to the direction of light $\mathbf{l} = [l_1, l_2, l_3]^t$. For a Lambertian surface, we can write the intensity of the image at a point $(x, y)_i$:

$$i(x, y)_i = i(x, y)_w = L_0 \rho(x, y)_w (\vec{\mathbf{n}}(x, y)_w . \vec{\mathbf{l}}). \tag{1.30}$$

It is a function of the intensity of light, the direction of light, the normal to the surface, and the reflection coefficient (albedo) $\rho(x, y)_w$ of the surface. If we take the logarithmic derivative of this function, we get:

$$d(\ln(i)) = \frac{d(L_0)}{L_0} + \frac{d(\rho)}{\rho} + \frac{d\vec{\mathbf{n}}.\vec{\mathbf{l}}}{(\vec{\mathbf{n}}.\vec{\mathbf{l}})} + \frac{\vec{\mathbf{n}}.d\vec{\mathbf{l}}}{(\vec{\mathbf{n}}.\vec{\mathbf{l}})}. \tag{1.31}$$

- The first term is linked to the *spatial variation of lighting* (which can be considered as constant over a small surface, except for cast shadows).
- The second term stands for the *nature of the surface*. If it is an uniformly painted surface, the spatial variations are small. If the surface is textured, the spatial variations are important. A low-pass spatial filtering of $\ln(i)$ will isolate the mean albedo of the surface and a high-pass filtering will select the texture. Because ρ is attached to the surface, the variations of the seen texture provide indications about the orientation of the surface (see shape from texture).
- The third and fourth terms indicate the *interaction between the direction of light* and the normal to the surface.
- The third term is related to the variations of the surface orientation (see shape from shading).

- The fourth term vanishes if the direction of light is constant over the scene, but it must be considered in the case of cast shadows.

1.4. From 2D Images to 3D Perception

There are several cues to 3D shape in a two-dimensional image; in particular we can infer such information from shading and texture, or from occlusions. The following 2D cues are related to the 3D structure of the scene:

- Occlusions
- Junctions of contours: in general, there are T-junctions when one object occludes another one, but there are X- or Y-junctions within the same object.
- Scaling or foreshortening of textures: the farther, the narrower.
- Parallel lines in perspective join at horizon.
- Intensities are stronger when the surfaces are oriented to the light.
- Direction of lightning: indicated by higher intensities and cast shadows.
- Cast shadows: a multiplicative process.
- Blur: objects in the distance are more blurred, and less bright.
- Motion is also an important means to recover shape, on some computational aspect, it is similar to stereo (two consecutive images may be considered as a stereo pair).

All these cues can be extracted from the distribution of gray (or color) levels in the image. Each of them, taken individually, can tell some information about the shapes and the relative positions of the objects in the scene, but they leave a certain amount of ambiguity.

In order to provide a reliable access to the scene structure, we should consider several cooperative processes between the 3D-feature extraction from these 2D cues. In fact, to solve this problem of shape from shading, our visual system (Gibson, 1950) often relies on prior assumptions, for example, that the illumination is from above or that the viewed surface is globally convex.

In the following section, we will give two examples showing how the shape of objects may be accessed from self-shading variations and from texture variations.

1.4.1. *Shape From Shading*

We should consider that extracting of shape from shading in real pictures remains a difficult problem. In order to partly solve this problem, we need to make a number of simplifications about the nature of the surfaces and about the lighting (Figure 1.27). In particular we will assume that:

- The source of light is unique and distant (the lighting of the scene is constant over the object surface).
- The surface of the considered object is Lambertian (constant albedo over the object surface, the surface is not textured).
- The surface is smooth.

Considering the world coordinates (we will drop the subscript w in this section), and under the above-mentioned restrictions, the image intensity is:

$$i(x, y) = L_0 \rho (\vec{\mathbf{n}}.\vec{\mathbf{l}}) = R(p, q),   \tag{1.32}$$

where (p, q) are variables in the so-called *gradient space*. A simple example is given in Figure 1.28.

The differential elements of world space and gradient space are related through the Hessian matrix of the surface equation $z = f(x, y)$. Remember that p and q are the first derivatives of the surface equation with respect to

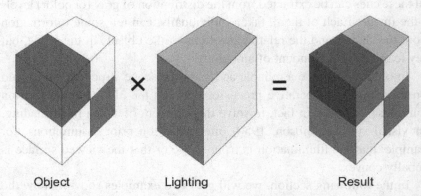

Object   Lighting   Result

Figure 1.27. The albedo of the object's surface is multiplied by the illumination of each face, which gives the observed image.

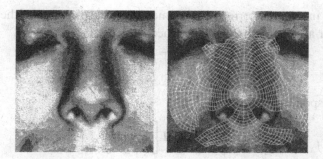

Figure 1.28. Recovery of shape from shading (Horn, 1986), reprinted with permission.

x and y:

$$dp = \frac{\partial p}{\partial x}dx + \frac{\partial p}{\partial y}dy = \frac{\partial^2 f}{\partial x^2}dx + \frac{\partial^2 f}{\partial x \partial y}dy,$$

$$dq = \frac{\partial q}{\partial x}dx + \frac{\partial q}{\partial y}dy = \frac{\partial^2 f}{\partial x \partial y}dx + \frac{\partial^2 f}{\partial y^2}dy$$

or in matrix form:

$$\begin{bmatrix} dp \\ dq \end{bmatrix} = \mathbf{H} \cdot \begin{bmatrix} dx \\ dy \end{bmatrix} \tag{1.33}$$

Let us calculate the gradient of the image intensity $i(x, y)$:

$$i_x = R_p \frac{dp}{dx} + R_q \frac{dq}{dx} = R_p \frac{\partial^2 f}{\partial x^2} + R_q \frac{\partial^2 f}{\partial x \partial y}$$

$$i_y = R_p \frac{dp}{dy} + R_q \frac{dq}{dy} = R_p \frac{\partial^2 f}{\partial x \partial y} + R_q \frac{\partial^2 f}{\partial y^2},$$

that is in matrix form:

$$\begin{bmatrix} i_x \\ i_y \end{bmatrix} = \mathbf{H} \cdot \begin{bmatrix} R_p \\ R_q \end{bmatrix}. \tag{1.34}$$

The image gradient is obtained from the gradient in the reflectance map by a multiplication by the Hessian matrix.

These equations have three unknowns $\partial^2 f/\partial x^2$, $\partial^2 f/\partial x \partial y$ and $\partial^2 f/\partial x^2$. In general this cannot be solved. However, the so-called "method of strips" can be applied. We start from a supposed known solution (x, y, z, p, q) and iterate the solution in a specific direction. Let us choose a path in the image that corresponds to moving in the direction of the steepest descent (or ascent) in the gradient space:

$$\begin{bmatrix} dx \\ dy \end{bmatrix} = \begin{bmatrix} R_p \\ R_q \end{bmatrix} d\varepsilon.$$

This corresponds to

$$\begin{bmatrix} dp \\ dq \end{bmatrix} = \mathbf{H} \begin{bmatrix} dx \\ dy \end{bmatrix} = \mathbf{H} \begin{bmatrix} R_p \\ R_q \end{bmatrix} d\varepsilon = \begin{bmatrix} i_x \\ i_y \end{bmatrix} d\varepsilon,$$

where $d\varepsilon$ is a small quantity. Hence, the step in gradient space is proportional to the step in image gradient.

In summary, the solution of five ordinary differential equations gives a curve on the object's surface:

$$\begin{cases} \dfrac{dx}{d\varepsilon} = R_p \quad \dfrac{dz}{d\varepsilon} = pR_p + qR_q \quad \dfrac{dp}{d\varepsilon} = i_x \\[3mm] \dfrac{dy}{d\varepsilon} = R_q \qquad\qquad\qquad\qquad \dfrac{dq}{d\varepsilon} = i_y \end{cases} \tag{1.35}$$

Hence, we require some starting values (x, y, z, p, q) for each strip; these may be related to special points on the surface, corresponding to maxima or minima in the reflectance map $R(p, q)$, or adjacent to the occluding contours. The procedure is as follows, starting at an initial point (x, y, z, p, q):

REPEAT   Take a small step $d\varepsilon$ along a strip from (x_0, y_0) to $(x_0 + dx, y_0 + dy)$.
This is parallel to the direction of the steepest ascent or descent in the reflectance map.
Compute changes in p and q from the changes in $i(x, y)$.
Define new values for p and q.
Compute new values for R, R_p and R_q.
Compute the new value for z
Reset start point to new value of (x, y, z, p, q).

UNTIL   a discontinuity is reached in the image function.

Note that, due to the formulation of the problem into differential equations, only the *relative depth* of a surface can be reached. In general, absolute depth is not reachable, because of the perspective projection, which drops one variable.

Caution: the perception of shape from shading has an inherent ambiguity due to the parity of the cosine function. We cannot tell whether the seen

Figure 1.29. Two examples where the variations in surface texture are an indication of the surface orientation.

surface is convex or concave. Our visual system is almost always biased toward convexity.

1.4.2. *Shape From Texture*

The definition of texture is not simple. The basic scheme is the periodic repetition of a local pattern over some area of an image. In fact, this pattern may not be constant and the repetition may be irregular (see Figure 1.29).

In general, the algorithms for shape from texture begin with a first step consisting of a local description of the texture projected on the image plane. The second step consists in the estimation of the spatial variation of this description. The last step interprets these variations in terms of the object's surface orientation.

There are several techniques to extract the information of texture variations, they are grouped in two large classes. The first one refers to statistical approaches, the second one refers to the spatial frequency approach.

In the statistical approach, textures are described in terms of *textels* (texture elements), then the statistics of the spatial distribution of the textels is computed. This approach is only valid with high densities of textels and fails in the case of uniform surfaces where only the borders are seen. Moreover, the description algorithms is often specific to each kind of texture.

The second approach does not require any assumption about the kind of texture involved. It is based on the description in terms of spatial frequencies,

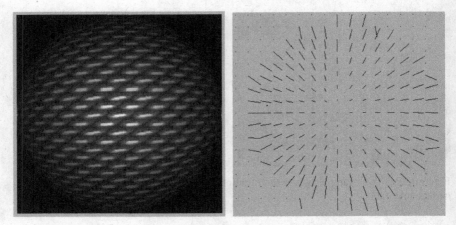

Figure 1.30. Textured surface of a sphere and projection of the surface's normal on the image plane, from El Ghadi and Guérin-Dugué (1999), with author's permission.

by means of Fourier or Wavelet transforms (Guérin-Dugué and Palagi, 1994; Sakai and Finkel, 1995; Super and Bovik, 1995; Malik *et al.*, 1997).

Super and Bovic (1995) use the second order moments of the local energy spectrum of spatial frequencies. Though restricting their model to an orthographic projection, they obtain interesting results. Guérin-Dugué and El Ghadi (1999) use an interesting estimation of texture spatial variations in terms of spatial *amplitude and frequency modulation* (the AM-FM model). Their results exhibit very good accuracy (Figure 1.30). In chapter 7, we present an approach based on a model of the visual cortex, which works with regular as well as irregular textures.

1.5. Conclusion

It is obvious that our final goal, which is "image understanding" from single or multiple 2D images, involves a number of very complex intermediate steps. In order to make a model of the structure of a scene, we ought to know both the physics of light (which states the transformations of incident light), and the opposite of the projective geometry which leads to image formation.

From a mathematical point of view, it is not a trivial task. Also, in order to keep tractable equations, we have to make many simplification hypotheses. Even in this restricted frame, the computational load remains important and often prohibiting in real-time applications.

In any case, it is interesting to consider that our visual system does not know the physics or the geometry of light, and that we still are able to grasp in one single glance a rather complete idea of a scene structure. It is known that it takes humans only 150 ms to detect the presence of a given object in a scene, and that it takes monkeys even less. Knowing that the brain of monkeys and even the human brain has trouble understanding mathematics. It is certain that we need not guess the above-mentioned algorithms in order to understand a picture. Maybe there are some other algorithms, with some ad-hoc neural implementations that are much more efficient. But they are designed for every day's life, and not for measurements and calibration purposes.

Chapter 2

The Visual System of Primates

This chapter is intended to give a "bird's eye view" about the biology of primates' visual system. Its general presentation is based on a classical scheme (anatomy and physiology). But, in each case when possible, special comments are given about the functional properties and the signal processing aspects of the considered neural structures.

2.1. Vision and Brain

2.1.1. *From Eye to Visual Cortex*

Frontal vision in primates implies a particular organization of the nerve fibers from eyes to brain. The brain consists of two hemispheres that are globally of similar organization but differ in some functional particularities.[1] Starting from each eye the optic nerve is separated into two parts: one part for the right hemifield of vision and the other part for the left hemifield.

For each eye (Figure 2.1), the right hemifield part of the optic nerve projects to the left hemisphere, and its left hemifield part projects to the right hemisphere. For that purpose, half of the nerve fibers issued from each eye follow a path to the ipsilateral hemisphere, but the other half must cross at a particular location called *Optic Chiasm* to reach the contralateral hemisphere.

---

[1]Roughly speaking, the right side is more devoted to space and global information, while the left side is more devoted to language and local information.

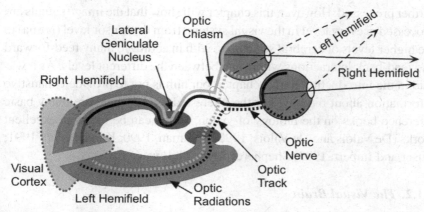

Figure 2.1. The visual path between the eye and the brain in species with frontal vision. As in the case of optics, where the projected images of objects are inverted (left-right, top-down), the information that is directed to each cerebral hemisphere is amazingly an inverted projection of left and right visual hemifields.

From this point, each optic track (the new name for the two halves of the optic nerves) reaches a part of the thalamus called Lateral Geniculate Nucleus (LGN). From there, the nerve cells issued from the LGN form the optic radiations that reach the first brain area: the primary visual cortex, or striate cortex, or area V1. Due to this organization, it is possible for each cerebral hemisphere to be locally informed of what is seen by each eye when both are looking at the same direction. Hence, stereoscopic information is available from the comparison between the images that are projected in each eye.

In order to make this comparison possible, the retinal topology must be preserved both at the LGN level and at the area V1 level, which is indeed observed. In vertebrates with lateral eyes (e.g. pigeons, rabbits, and preys in general) or with independent eyes (horse fish, chameleon), the two parts of the brain process two different independent images and there is no need to compare them on a topological basis. It is noteworthy that predators are generally endowed with frontal vision, which allows an accurate estimation of the prey's position whereas preys are rather endowed with a wide field lateral vision, which allows them to detect danger more easily.

At this stage of presentation, a naive reader could think that the eyes behave merely like cameras that send to the brain the information to be further processed. However, this chapter will show that the image signals are processed at every level in the visual system, from the sensor level (retina) to the higher levels of cerebral structures, with in addition, many feed-forward and feed-back interactions within and between the different levels. As it was said in the introduction to this chapter, our aim is not to provide exhaustive information about biology. For the interested reader, there are many basic reference books on the biology of vision, which can be found in excellent works (De Valois and De Valois, 1990; Spillman, 1990; Kandel *et al.*, 1991; Buser and Imbert, 1992; Shepherd, 1994; McIlwain, 1996).

2.1.2. *The Visual Brain*

From the entry of optic radiations, the cortical surface devoted to vision is about 15% of the whole brain for humans, and 50% for monkeys. In fact there are some well identified visual cortical areas starting at the occipital part of the brain, but by means of associative areas, almost the whole brain may be concerned at one moment by vision. Figure 2.2 gives a side view of the brain with the four main regions (lobes), occipital, parietal, temporal

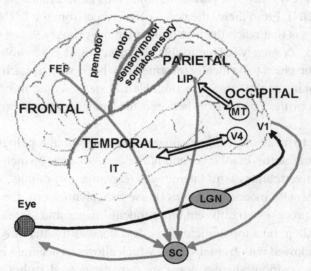

Figure 2.2. The eye and the different regions of the brain concerned with vision. V1: primary visual area, MT: area in relation with motion, V4: area in relation with forms, V4-IT: ventral stream (linked to objects, "What"), MT-LIP: dorsal stream (linked to action, "Where").

and frontal lobes:

- The occipital lobe is the main entry point and the main region which is specific of visual information.
- The parietal lobe is a large associative region concerned, in the right hemisphere, with the manipulation of objects in space (taking, touching, or seeing). In the left hemisphere it is more concerned with words that are spoken, read or even touched (in the case of blind people). It is also involved in long-term reflexive memory.[2]
- The temporal lobe is more involved in long-term declarative[3] memory, pictorial memory for the right hemisphere, and verbal memory for the left hemisphere.
- The frontal lobe is concerned with short-term working memory that stores the location of objects in the absence of explicit cues, or that plans the sequence to realize a movement.

The superior colliculus (see bottom of Figure 2.2) is a structure involved in the control of eye movements. About 10–20% of the optic nerve's fibers is routed to the superior colliculus (SC), making it aware of *where* something of possible interest is. As a reflex the superior colliculus sends an order to orient the eyes toward the object. This constitutes a first loop Eye-SC-Eye. The remaining 80–90% of the fibers reach the lateral geniculate nucleus (LGN) and from there is routed to the primary visual cortex V1. The superior colliculus receives also a feedback from the primary visual area, which tells where, in the visual field, a texture, or a color is different from its surroundings, initiating a movement of the eye toward this location. This constitutes a second loop: Eye-LGN-V1-SC-Eye. Further, the lateral intraparietal area (LIP) sends a more elaborated information to SC, for example in order to verify if an object out of the visual field is still here. This is the third loop: Eye-LGN-V1-LIP-SC-Eye. All these three loops of controls to SC work unconsciously. The only conscious control is sent by the frontal eye field (FEF) to SC to make a voluntary saccade in order to explore a given position or to do a pursuit movement in order to follow a moving object.

In Figure 2.2 the occipital lobe sends two tracks, one through area MT towards the parietal cortex, and one through area V4 towards the

---

[2]Or procedural memory: Knowing (unconsciously) *how* to do something.

[3]Knowing (consciously) *that* an object exists or *that* something has happened.

inferotemporal cortex. The first pathway is related to location in conjunction with motion or intended movements. It has been called the "Where" pathway. The second one is more concerned with the forms and colors in conjunction with object recognition; it has been called the "What" pathway.

This classification has been very successful because of its simplicity. But it is now widely recognized that things are more complex: the information is not so well segregated between the two pathways and they are linked through many associative connections.

Now, knowing that things are not as simple as they look, we are ready for a more in-depth analysis of the visual system. Although it should be clear that this presentation has been simplified for non-biologists to understand it, important remarks about the processing of signals at every level still need to be made.

2.2. Primates' Eyes

2.2.1. *Geometry of the Eye*

The geometry of the eye's optics is very well known (Atchison and Smith, 2000). It allows to compute the exact projection of the surrounding world on the retina:

The refracting power of the eye is due to curved surfaces, which separate media with different indices of refraction. The most significant ones delimit the cornea and the crystalline lens.

- tear film 1.34
- cornea 1.37
- aqueous humor 1.33
- crystalline lens 1.42
- vitreous humor 1.33.

The refracting power of the cornea is between 39 and 48 diopters, while the one of the crystalline lens is less, between 15 and 24 diopters. The power of this system is so strong that the second nodal point is just behind the lens inside the vitreous humor. The retinal image is roughly spherical, with its center being this nodal point. This fact has several consequences as will be explained at the next section.

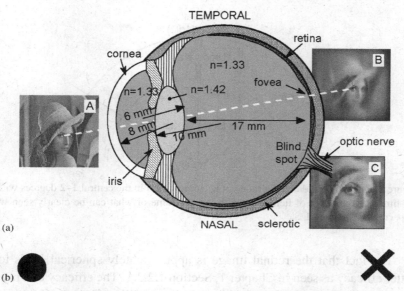

(a)

(b)

Figure 2.3. (a) Geometry and optics of the eye and retinal projection of images. A: observed image, B: its projection on the retinal surface, C: the image that would be carried by the optic nerve due to the variable density of photoreceptors. (b) Get conscious of your blind spot: close your right eye and focus on the cross, then, by moving back and forth, you will find a distance that makes the black disc disappear: this is when its projection is on your blind spot.

The point where the optic nerve leaves the eye is called the "blind spot", its size is the one of an orange at arm's length. To "see" your blind spot, look at Figure 2.3 and follow the instructions.

2.2.2. *Functional Consequences*

Because the crystalline lens is a thick lens, the image is precisely focused on the retina only near the optical axis: its periphery is blurred (Figure 2.3, insert B). In accordance with this property, the density of the photoreceptors on the retina's surface is maximal at the center and decreases progressively toward the periphery (see later). Figure 2.3 illustrates this fact. If the photoreceptors were re-arranged on a regular grid while keeping their image samples, the resulting image (Figure 2.3, insert C) would strongly magnify the center. Figure 2.4 illustrates the size of accurate vision (~1 deg) in relation to the whole field of vision (160 deg).

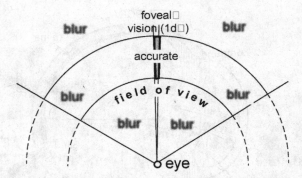

Figure 2.4.   Visual field and the range of accurate vision in the central 1–2 degrees within the limits of the depth of field. Notice the tiny volume of what can be clearly seen with respect to the 3D world.

The fact that the retinal image is approximately spherical leads to a better efficacy as seen in Chapter 1, Section 1.2.3.1. The efficacy of a ray of light making an angle θ with the optical axis falls off as $\cos(\theta)$ instead of $\cos^3(\theta)$ for usual photo cameras. Another property of the spherical projection is a kind of compression of coordinates: if an object lies on a plane at a distance X from the optical axis, the plane being itself at a distance Z from the eye, the projection of this object on the retina is at an angle $\theta = \arctan(X/Z)$.

The retinal image depends on various active factors that are more or less automatically controlled:

- Its luminosity is controlled by the iris diameter,
- Its focus is realized by mean of cilliary muscles which modify the curvature of the crystalline lens (accommodation),
- Its position and orientation are controlled by extrinsic muscles attached on the choroid.

While looking without any point of fixation, the eye accommodates for a distance of 1 meter. Accommodation is a reflex movement which latency is 300 ms. Both eyes accommodate simultaneously.

Because the range of accurate vision is very small as compared to the surrounding world, there is an imperative need to move the eyes in order to explore the visual environment. For this, two pairs of muscles allow left-right and up-down movements. A third pair allows a rotation around the optical axis (cyclotorsion movement).

There are three kinds of large eye movements:

- Saccades: they rotate both eyes and bring the image of interest onto the fovea, for example when reading this book. Due to the very low inertia of the eye, saccades may be very fast, up to $600\,d°$ per second (that is a jet seen at 20 m and flying at a speed of 600 kmh).
- Vergence: it is an eye movement caused by looking at different objects at different distances. We talk about Convergence or Divergence, depending on the fact that we look from far to near or from near to far.
- Pursuit: if an object that we are looking at moves slowly, a pursuit movement is generated in order to keep its image still on the retina.

There are also three kinds of miniature eye movements:

- Drift movements: they are slow, with amplitudes of around 2–5 minutes and velocities around 1 minute of arc per second. Drift movements of each eye are not correlated.
- Microsaccades: they have the same function as the large saccades. They are primarily corrective, they occur at the end of each drift movement.
- Tremor: the amplitude of tremor is very small, about 5-10 seconds of arc, not far from intercone distance.

Other movements exist, especially during head rotation or when a significant part of the field of vision moves: vestibular ocular reflex or optokinetic reflex occur, tending toward the stabilization of the retinal image.

2.3. The Retina

The retina is the innermost layer of the eye, where the incoming light is focused (Figure 2.5). It is composed of nerve cells, which sense and process the 2D image of 3D world. The optic nerve is the output of the retina, it sends impulses to the brain, which will further analyze and interpret this image.

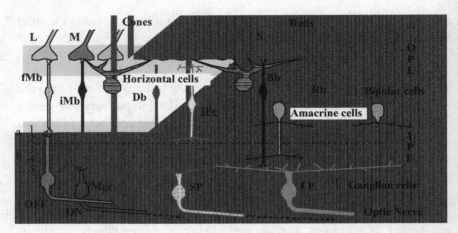

Figure 2.5. The retinal cells. The left-hand side represents the central retina (fovea) and the rightward direction represents increasing eccentricity. OPL: Outer Plexiform Layer. L, M, S: types of cone photoreceptors. fMb, iMb: flat and invaginating Midget bipolar cells. Db, Bb, Rb: diffuse, blue cone, rod bipolar cells. IPL: Inner Plexiform Layer. Mgc, SP, LP: Midget, small parasol, large parasol ganglion cells.

The shape of the human retina is roughly spherical (with a radius of 17 mm) and its area is about 1100 mm^2. Unfolded, it would be a disc of 4.2 cm in diameter. At the center one degree of visual angle is equal to 290 μm on the retina.

Between the choroid and the vitreous humor, 10 different layers compose the retina. Among them, six are of functional importance for image processing:

- The outer nuclear layer where cell bodies of photoreceptors are found,
- The outer plexiform layer (OPL) where photoreceptors, horizontal cells, and bipolar cells make contact, that is where the first level of processing takes place.
- The inner nuclear layer where the somas of horizontal, bipolar, and amacrine cells, as well as the Müller cells lie.
- The inner plexiform layer (IPL) where bipolar, amacrine and ganglion cells are interconnected, that is where the second level of processing takes place.
- The ganglion cells layer where the bodies of ganglion cells and some amacrine cells are found.

- The axons of the ganglion cells, which further traverse the retina at the optic disk (blind spot).

The incoming light passes through all these layers before reaching the photoreceptors. Light causes a cascade of chemical reactions with visual pigments and other molecules, resulting in the variation of the electrical potential of the photoreceptor's membrane. Activated photoreceptors in turn stimulate the other retinal cells all the way to the axons of the ganglion cells, which propagate the processed information to the brain through the optic nerve.

At the center of the retina, the axons of the nerve cells are drawn laterally, which produces a region of reduced thickness: the fovea ("pit" in Latin). The cross diameter of the central fovea from rim to rim is 1.5 mm ($\sim5°$).

2.3.1. *Photoreceptors*

There are two categories of photoreceptors: cones (about 6.5 to 7 million in each eye) and rods (about 120 to 130 million in each eye). The former ones are dedicated to photopic vision (day light) and colors, the latter are active in scotopic vision (dim light) and do not allow color vision.

There are 3 types of cones that are identified by their spectral sensitivity curves (Smith and Pokorny, 1975; Schnapf *et al.*, 1988; Jacobs, 1996b): L, M and S for Long (560–565 nm), Medium (535–540 nm) and Short (430–440 nm) wavelengths respectively (see Chapter 5 for more details). It should be noted that there is a significant overlap between L and M cones spectra, whereas the spectrum of S cones is shifted more toward violet, an important functional consequence of this fact will be explained in Chapter 5.

It should be noted that trichromacy is not a rule: at least four visual pigments are known in several species like turtles, alligators, lizards or birds, showing that tetrachromacy is a possible code and why not "pentachromacy"...

2.3.1.1. *Spatial Sampling*

Cones and rods are more or less regularly arranged (Marc and Sperling, 1977; Maloney, 1996; Roorda and Williams, 1998) at the surface of the retina. They fit locally a centered hexagonal grid (which represents a 15% increase of efficacy with respect to a square grid).

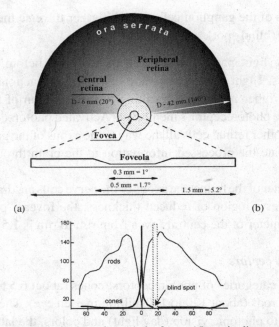

Figure 2.6.   Photoreceptors in the retina. (a) the fovea, 15 mm in diameter (5.2° of visual angle), contains mostly cones and very few rods. The foveola, circa 0.3 mm and 1° of visual angle, is rod free and has no S (blue) cones. The parafovea contains less and less cones and more rods, which in turn decrease in density toward the periphery. (b) spatial distribution of photoreceptors (after Osterberg, 1935, reprinted with permission).

Their density is far from being uniform on the retinal surface (Figure 2.6). The cones are mostly present at the center with a density of 180,000/mm^2, which rapidly decreases with eccentricity.[4] The rods, absent from the center increase progressively in density to reach a maximum of 160,000/mm^2 at 18° eccentricity.

At the center of the fovea there is a smaller area: the foveola ($\sim$300 μ in diameter) where the rods are absent. That is the reason why you cannot see a small star at night when looking directly at it, but you will see it when looking next to it. S cones are missing also in this area.[5]

---

[4] This density gives a mean spacing of 2.5 μ between cones. As one degree is 290 μ, this gives about 120 cones per degree at the center of the fovea, which is compatible with our maximum visual acuity of 60 cycles per degree.

[5] This can be verified on your computer screen by putting a blue pixel on a yellow background: when looking at the pixel it appears to be black, but it becomes blue when looking next to it!

2.3.1.2. *Function*

The photoreceptors convert light in a highly efficient fashion[6] into an electric signal by means of a cascade of chemical reactions starting with photopigments (rhodopsin for rods and three types of opsins for cones). These reactions will be analyzed in details later (see Chapter 6 about non-linearity). At rest, the cellular membrane of a photoreceptor is depolarized, the incoming light turns it into a hyperpolarization.

The output end of the cone cell is known as the pedicle, and that of the rod cell as the spherule. Both synaptic endings are filled with synaptic vesicles, which means that photoreceptors' information is transmitted through a chemical synapse to target cells. The neurotransmitter is known to be glutamate.

Two important properties of photoreceptors are worth mentioning: their spatial coupling and their ability to adapt to light intensity.

Spatial coupling: Neighboring photoreceptors are coupled by electrical synapses called "Gap Junctions" (Attwell *et al.*, 1984): by this mean, a partial sharing of their internal potential, the photoreceptors' output image is a slightly smoothed version of the retinal image. As it will be seen later, this reduces the spatial noise due to mismatches among receptors. It is postulated that only receptors of the same type (rods, L-cones, M-cones or S-cones) would be coupled. Even if this hypothesis is not fully confirmed in the case of primates, we will show that it may be of some importance for color processing in the retina.

Adaptation: The range of light conversion by photoreceptors is limited to 1.5 decade. Because the range of light arriving on it is of the order of

Table 2.1. Comparison between cones and rods.

	Cones	Rods
Sensitivity	Low	High
Saturation	Only for intense light	In daylight
Temporal response	Fast	Slow
Better for	Axial rays of light	Scattered light

---

[6]In rods, the transmembrane potential shows variations which correspond to individual photons.

6 decades, a photoreceptor adapts to the mean ambient Lighting, coding for 1.5 decade around this mean Lighting (more details are given in Chapter 6.

2.3.2. *The Outer Plexiform Layer*

2.3.2.1. *Synaptic triad*

The synaptic triad (Boycott *et al.*, 1987; Dowling, 1987) is the functional unit of the outer plexiform layer (Figure 2.7): it is the location where photoreceptors, bipolar cells and horizontal cells interact. The simplest structure is in the fovea where a bipolar cell connects only one photoreceptor. It consists typically of an invagination of the cone pedicle, or of the rod spherule, where the dendrites of second order neurons are found:

- the dendritic terminal of an "invaginating bipolar cell", as center element,
- the dendritic terminals of horizontal cells, as two (or more) lateral elements.

A third connection is present at the base of the cone pedicle: the dendrite of a "flat bipolar cell". There are no such flat junctions on rod spherules.

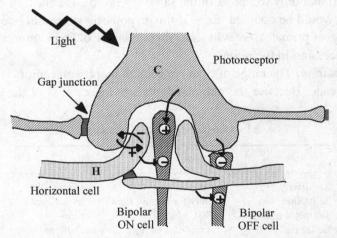

Figure 2.7.   The synaptic triad: the light arriving in the photoreceptor (C) excites the ON bipolar cell and inhibits the OFF bipolar cell. The horizontal cells, coupled by gap-junctions, operate a lateral inhibition process, either by feedback on the receptor, or by feedforward on bipolar cells.

We have said that the photoreceptor is depolarized in the dark, conversely it is hyperpolarized by the incoming light. The cone-to-invaginating bipolar cell synapse is sign-inverting, which makes incoming light depolarize the bipolar cell membrane. Since the cell is excited as the light is turned on, it is called an "ON bipolar cell".

The cone-to-flat bipolar cell synapse is sign preserving, which makes the incoming light hyperpolarize the bipolar cell membrane. But when the light is turned off, the cell membrane is then depolarized. Because this cell is excited when light is turned off, it is called an "OFF bipolar cell" (see Werblin and Dowling, 1969; Werblin, 1991).

The cone-to-horizontal cell synapse is sign preserving: horizontal cells are hyperpolarized by light. Horizontal cells make a negative feedback loop to photoreceptors so that their action is opposed to that of the light. Horizontal cells also send a feed-forward negative signal to bipolar cells with an action also opposed to that of light.

Moreoverver, as well as the receptors, the horizontal cells are linked by gap junctions: they share a fraction of their membrane potential with their neighbors. This way, when a small spot of light is present on a photoreceptor, the signal spreads into the horizontal cells network, and negative actions can be sent far from the point of stimulation, well beyond the cells' dendritic trees.

In other words, a photoreceptor (a bipolar cell) receives a negative feedback (feed-forward) signal from the network of horizontal cells, which extends over a wide spatial area. This action is in fact the origin of the processing of visual signals by the OPL: from an input signal (cones), we subtract a smoothed version of itself, this produces the shape of the receptor field of bipolar cells (Figure 2.8), the contrast enhancement effect that is observed in the Mach bands phenomenon.[7] Because of the center/surround opposition, we speak of "ON center-OFF surround" bipolar cells, or conversely of "OFF center-ON surround" cells.

In terms of signal processing, this property is called "high-pass" filtering, that is the removal of low-frequency components: the uniform regions of the signal are damped, so that only the "edges" seem to be preserved.

---

[7] At the junction between two grey surfaces, the darker one looks even darker near the boarder, and the lighter one looks clearer on the opposite side of the boarder.

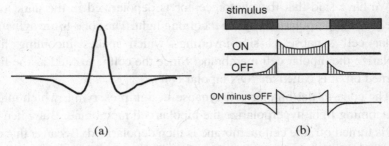

Figure 2.8. Bipolar cells response. (a) cell's receptor field. (b) response of ON and OFF-type cells to a bar of light on a gray background (top). Equivalent signal by taking the difference between ON and OFF cells.

In fact, the processing which is operated here is a little more complicated because the cells have a temporal response. It results in a spatio-temporal behavior of great functional importance, which will be studied in detail in Chapter 3 "Basic model of the retina".

We should note that all the cells in the OPL transmit only graded information (concentrations of neurotransmitter at the chemical synapses and graded potentials at the gap junctions) but no nerve impulse. In other words, the OPL is a place where "analog" signals are processed and transmitted.

2.3.2.2. *Bipolar Cells*

The bipolar cells constitute the output of the first processing stage (OPL). At least 10 types of bipolar cells have been identified in the retina:

- flat and invaginating "Midget" bipolar cells (fMb and iMb) are mainly found in the fovea, they connect only one cone and only one ganglion cell (Wässel and Boycott, 1991).
- diffuse bipolar cells (DB) connect several cones in OPL and several ganglion cells in the IPL. There are six types named from 1 to 6 according to the extension of their dendritic trees. Low numbered types are more represented in central retina and connect 5–7 cones, high numbered types are more and more numerous with increasing eccentricity and connect 12–15 cones.
- rod bipolar cells (Rb) are of the ON type and are specific of the rod circuit.

- S-cones (or blue-cones) bipolar cells (Bb) are also specific for this photoreceptor. Their role will be studied later.
- giant bistratified bipolar cells (GBb) have a 70–100 μ dendritic tree connecting 15–20 cones, and project endings in the two strata of the inner plexiform layer.

For each cone both types of ON and OFF bipolar cells are present. As the ON cell receives a {Cone minus Horizontal cell} signal and the OFF cell receives a {Horizontal cell minus Cone} signal, it can be considered that the fictive path {ON minus OFF} carries the difference between cone and horizontal cell signal, regardless of its sign. It can be argued that ON and OFF paths exist only because nerve cells cannot transmit negative signals, i.e. negative concentrations of neurotransmitter. We will see later that the separation into two paths (ON and OFF) is in fact useful for some processing purposes.

Three particular important facts should be mentioned here:

- The circuit for the S-cones is particular (no OFF bipolar cell, specific horizontal cell type). This has some implication for color processing, which will be studied in depth in chapter 5 "Color coding in the retina".
- Because rod spherules have no flat junction, rod bipolar cells are only of ON type. The separation into ON and OFF pathways takes place in the interplexiform layer where AII amacrine cells play a role similar to the OPL's horizontal cells.
- Bipolar cells, mainly those of the OFF type, are connected through gap junctions (Hare and Owen, 1990). Such junctions may cause bipolar cell receptive fields to resemble those of horizontal cells, spreading far beyond their dendritic tree.

2.3.2.3. *Horizontal Cells*

Horizontal cells, being coupled through gap junctions, make a functional network that can be seen as the structure of a recursive low-pass spatial filter. Space constants of cat horizontal cell receptive fields range from 200 to 450 μm. Dacheux and Raviola (1982) suggest that this form of coupling responds well to a principle of "economy": with a connection to the first neighbor, this network provides receptive fields much larger than the dendritic extension.

Horizontal cells are known to play a role of lateral inhibition in the OPL, either by negative feedback on the photoreceptors (Kamermans and Spekreise, 1999), or (and) by a feedforward inhibition to bipolar cells (Yang and Wu, 1991; Hare and Owen, 1992). In the feedback case, two types of actions are possible and might coexist: either the feedback is said to be "$GABA_A$-ergic" (O'Bryan, 1973; Kaneko and Tachibana, 1986) and applies to the signal itself, resulting in a high-pass behavior for the OPL, or it acts on the calcium current of the photoreceptor (Verweij *et al.*, 1996) and leads to a modulation process that controls the photoreceptor's gain. Such an action has been postulated as a good candidate for the color constancy effect (Kamermans and Speckreise, 1999). A similar effect of gain control has been described for bipolar cells (Fahey and Burkhardt, 2003).

For example in the case of fish, retina horizontal cells' dendrites connect to cones as inhibitory feedback terminals, which results in a high-pass behavior of the OPL. Moreover, it has been established that this interaction is able of plasticity: while suppressed in the dark, feedback is potentiated by light adaptation (Djamgoz and Kolb, 1993). This modulation, which is light evoked and does not appear to be mediated by a GABA receptor, causes a change in the rate of transmitter release from the cone with no change in the cone's membrane potential. In mammalian cones some kind of feedback is also expected at this site but adaptation to light has not yet been shown (Packer and Dacey, 2001).

It will be shown in Chapter 3 that a combination of feedback and feedforward inhibition by horizontal cells is able to optimize the frequency response of the OPL.

There are two main types of horizontal cells named H1 and H2 (Kolb *et al.*, 1980). Both cell types give the same slow hyperpolarizing response to a light flash, whatever the wavelength of light is. H1 cells are sensitive to L- and M-cones stimulation, but are insensitive to S-cones stimulation. H2 cells are very sensitive to the S-cones stimulation and their responses are the same for S- (blue), L- (red) and M- (green) cones (Dacey *et al.*, 1996). In conclusion, the primate's horizontal cells are achromatic, their response is of same sign whatever the wavelength. On the contrary, in fish and turtle retina, the horizontal cells respond differently to different cone stimulations. For many biologists a question is still not answered: which trick can lead achromatic horizontal cells to chromatic opponent bipolar and ganglion cells? In Chapter 5 we give a simple model which can account for this property.

2.3.3. *The Inner Plexiform Layer*

Here takes place the second processing level of the retina. It implies that the bipolar cells, the amacrine cells and the ganglion cells, the axons of which constitute the optic nerve. It is only at this level that nerve impulses appear (in ganglion cells and in some amacrine cells). We should keep in mind that all the other cells transmit only graded potentials. The processing of signals here look like the one of the OPL, but with more complexity and with the addition of a temporal component.

2.3.3.1. *Amacrine Cells*

Amacrine cells are interneurons that interact at the level of the bipolar-ganglion cell junction. There are 25 known types of such cells. They make synapses in the IPL and are known to add a temporal aspect to the bipolar cell message presented to the ganglion cell. They also play a role in the modulation of activities of other retinal cells.

Amacrine cell synapses are frequently seen to be reciprocal to bipolar input, i.e. the amacrine returns a synapse in the vicinity of the input synapse. As they are inhibitory neurons, this reciprocal synapse adds a transient behavior to the bipolar signal. Their output also concerns other amacrine cells and ganglion cells.

Some examples:

- the A2 cells receive bipolar input from OFF-center cone bipolar cell and make reciprocal synapses with them. They also make an inhibitory synapse upon the corresponding OFF-center ganglion cell.
- The AII and the A17 amacrine cells play the same role for rod signals.
- The A8 amacrine cell is involved in the cone pathways with a rather transient OFF-response to light.
- A13 cells make reciprocal synapses upon cone bipolar cells and possibly upon rod bipolar cells. A13 makes synaptic output to OFF-center ganglion cells.

Three types of amacrine cells are worth mentioning because of their particular role in signal processing.

Transient ON-OFF cells: A19 and A20 are both transient depolarizing ON-OFF cells otherwise known as ON-OFF amacrine cells in response type. For a long time, it has been difficult to record responses from such cells in mammalian retina (Freed *et al.*, 1996). They receive inputs from OFF-center and ON-center cone bipolar cells. We will see in the next chapter that the interest of such cells may be to tell "where" and "when" something happens. Gap junctions may interconnect dendrites of A19 cells.

Dopaminergic cells: A18 (or Type 1CA) are dopaminergic amacrine cell types, with a wide dendritic field in stratum 1 of the IPL, making synapses on amacrine cells. Their dendrites emit long axon-like processes touching different strata of the IPL and the ganglion cell layer. Sometimes they also reach the OPL, giving rise to an interesting process of modulation on the horizontal cells function: Dopamine cells in the fish retina are interplexiform cells (IPC) known to modulate the spatial extent of horizontal cells by uncoupling their gap junctions (Teranishi *et al.*, 1983). In mammals a similar activity, but with weaker effect, has been reported (Pflug and Nelson, 1989; Bloomfield, 1993).

The ACh amacrine starburst cells: occurring as symmetric pairs, may be involved in direction selectivity in the rabbit's retina. In the salamander retina, some direction selectivity has also been described (Werblin *et al.*, 1988). Similar cells exist in the human retina but with less clear branching pattern. And direction selectivity has not been reported up to now.

2.3.3.2. *Ganglion Cells*

There are three kinds of ganglion cells which are identified by Greek letters (α, β, γ) in a monkey's retina, and roman letters (X, Y, W) in a cat's retina, they are sometimes identified by their target cells in the lateral geniculate nucleus. In this case, they are said to belong to the Parvo- Magno- or Koniocellular pathway.

The *midget* ganglion cells are directly connected in a one-to-one manner to the midget bipolar cells. They are of ON and OFF types and are said to be of X type in a cat's retina, of β type in a monkey's retina, they belong to the *parvocellular* pathway because of their small target cells in LGN. They carry wide band signals, they are high-pass in spatial frequency, low-pass in temporal frequency and their response exhibits inseparability of time and space variables (Beaudot *et al.*, 1993). They transmit color signals with chromatic opposition: Red/Green or Blue/Yellow (De Monasterio and Gouras, 1975), just as the corresponding geniculate cells (De Valois *et al.*, 1966). However, a given midget cell also responds to luminance information and to chromatic information.

The *small parasol* and *large parasol* ganglion cells belong to the *magnocellular* pathway, they are of Y type in a cat's retina and of α type in a monkey's one. Their response concerns the low spatial frequencies of the retinal image, with a temporal aspect markedly transient, probably due to the interaction with amacrine cells (Jacobs and Werblin, 1998). A group of these cells is of type either ON or OFF with a linear response (the response to the sum of two stimuli is equal to the sum of the individual responses to each stimulus). They carry mainly luminance signals, they are thought to be implied in the perception of motion. A second group presents a highly non-linear, transient and spatially low-pass response, the cells are of ON-OFF type, that is, they respond as well to the apparition of light as to its removal. Hence they appear to be a signal of WHERE and WHEN something happens in the visual field. Many of these cells are directed to the superior colliculus. They also carry luminance information.

A third kind of cells, targeting the very small cells of the LGN (*Koniocellular* pathway), are not fully understood, they present various characteristics of the previously defined types, they are called γ in the monkey's retina and W for the cat. 90% of these cells project to the LGN, the remaining 10% project to the superior colliculus. It has been recently discovered that the blue-yellow ganglion cells project to the koniocellular layers of the LGN.

Because the rod bipolar cells are only of ON type, the OFF component is generated by means of the AII amacrine cells. The rod OFF signals result from an inhibitory synapse between AII amacrine cells and both OFF cone bipolar cells and OFF ganglion cells at the level of sublamina *a* of the IPL.

Among the ganglion cells, about 80% form the parvocellular pathway; 10% form the magnocellular pathway; the remaining 10% constitute the Koniocellular pathway.

2.3.3.3. *Interplexiform Cells*

In every species there are cells that feed information from the IPL back to the OPL: the interplexiform cells. In cat, monkey and human retinas such interplexiform cells get input from amacrine cells and make synapses upon rod and cone bipolar cells (Kolb and West, 1977; Nakamura *et al.*, 1980; Lee *et al.*, 2003). In the fish retina the interplexiform cells are dopaminergic, they make synapses in the OPL, primarily upon horizontal cells. Dopamine is known to modify the coupling between horizontal cells (Witkovsky *et al.*, 1989) and also to be involved in the day-night cycle. It is observed that this kind of feedback modulates the spatiotemporal transfer function of the OPL according to Lighting: the low-pass filtering is stronger under weak illumination, which may result in increasing the signal-to-noise ratio.

Let us remember that in mammals the A18 amacrine cells can play the role of interplexiform cells. As they are dopaminergic and, in human retina, make synapses with horizontal cells, they may play the same role as the IPC's in fish retina.

2.3.4. *Summary: Output Signals of the Retina*

It seems that first of all the Y ON-OFF cells tell the brain where and when something is happening in the visual field. Secondly, the ON and OFF Y cells send a gross signal giving the outlines of the visual scene. Then the signal of the X cells, providing the missing details arrives. This scheme might be of interest in a kind of "coarse-to-fine" process.

After a first stage of processing, the retina divides the information into several streams that reach the brain at different time slots.

2.4. Lateral Geniculate Nucleus

2.4.1. *Anatomy*

There is some segregation of the two types of receptive fields in the LGN. Although layers 1 and 2 each contain a mixture cells with ON/OFF and OFF/ON receptive fields, layers 3–6 contain only one or the other type of receptive field (two layers contain only ON/OFF and two contain only OFF/ON).

Table 2.2. Summary of the retinal outputs.

Retina			Type of cell	LGN	Properties
β	X	ON	Midget	Parvo	High spatial frequency
					Sustained
		OFF			Chromatic opponent
					Slow
α	Y	ON	Small and large	Magno	Low spatial frequency
		OFF	parasol		Transient
		ON-OFF			Achromatic
					Fast
γ	W	≈	≈	Konio	Between parvo and magno?

The surrounding of LGN cells has a stronger influence than that of ganglion cells. This means that the surrounding in LGN receptive fields is more weighted, and that, consequently, contrast information is amplified.

The Lateral Geniculate Nucleus (LGN) is composed of six layers of cells (Figure 2.9) numbered from 1 (ventral) to 6 (dorsal).

Layers 1 and 2 are called "Magnocellular" because they made of large nerve cells. They are the targets of the axons of parasol ganglion cells (α or Y cells) and they project to the primary visual cortex in layer 4Cα. The pathway from the retina to the primary visual cortex that passes through these cells is called the "M pathway".

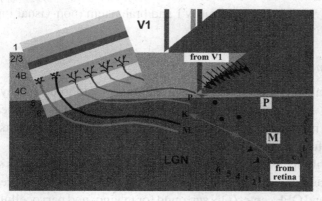

Figure 2.9. Lateral Geniculate nucleus. Six layers for magno- and parvocellular streams, with alternative representations of ipsi- and contralateral eye. The retinal topography is preserved up to the V1 cortex where signals from both eyes are combined for stereoscopic vision.

Layers 3 to 6 are called "Parvocellular" because they are made of smaller cells. They are targeted by the midget ganglion cells (β or X cells) and project to layer $4C\beta$ of the primary visual cortex. The retinocortical pathway that passes through these cells is called the "P pathway".

Each layer receives afferences only from one eye. The nasal fibers from the contralateral retina project to layers 1, 4 and 6. The fibers from the temporal ipsilateral retina (same side of visual field) project to layers 2, 3 and 5. At this level there is no convergence of axons of both retinas: the pathways are monocular.

In each layer the projection of the visual hemifield of one eye is retinotopic. The maps are anatomically overlaid so that a point in the visual field corresponds to a projection column perpendicular to the layers.

There is a third pathway. Retinal γ or W ganglion cells project to very small cells situated between the LGN layers. Because of the size of the target cells, it constitutes the Koniocellular or K pathway. The K cells send axons to the layer III cytochrome oxidase blobs, and to layer I of the primary visual cortex.

V1 targets of M and P pathways are usually considered to be simple cells (Sciller and Malpei, 1978). Recently, V1 targets of the three pathways have been observed in owl monkeys. It appears that V1 cells postsynaptic to K and M axons exhibit orientation selectivity, whereas P axons' postsynaptic cells do not (Shostak *et al.*, 2003).

In fact, the retinal cells constitute only 10% of the LGN inputs. The remaining 90% are issued from V1 and brain stem (non-visual inputs).

2.4.2. *Function*

The Lateral Geniculate Nucleus is traditionally said to be a relay between the retina and the primary visual cortical area V1. It now appears as an important node in the processing of visual information, even if its function is not yet fully understood.

Roughly speaking, the three pathways of the LGN present the same spatial characteristics as the retinal ganglion cells they receive: ON center/OFF surround and OFF center/ON surround for magno- and parvocellular layers, ON-OFF for magnocellular layer, and more variable for the K pathway (Schiller and Malpei, 1978). All cells are monocular.

Magnocellular pathway: cells' receptive fields are 2–3 times larger than parvocellular cell receptive fields. They have better sensitivity to contrast; they are faster and respond well to temporal signals and to moving stimuli. Their temporal response is transient.

Parvocellular pathway: cells have a better spatial acuity, their temporal response is sustained, and they respond to color stimuli by chromatic oppositions (Blue-Yellow and Red-Green).

Koniocellular pathway: it is a heterogeneous group. At any eccentricity, the temporal contrast sensitivity of K cells lies between that of M and P cells. Response properties of K cells are more similar to those of P cells than those of M cells. Some K cells in primates respond to color Blue-ON cells, (Casagrande, 1999). In some studies, K cells have proven to have the widest receptive field among P and M cells (Xu *et al.*, 2002).

Among the possible functions of the LGN, one is thought to be the enhancement of information about contrast. But note that 80 to 90% of the LGN inputs do not originate in the retina: a modulation of its processing with arousal is possible via the reticular activating system, A modulation is also possible via the feedback from V1.

Another fact that may be of importance is that there exist two kinds of cells, which are temporally different: lagged and non-lagged cells (Mastronade, 1987; Cai *et al.*, 1997). On the spatial point of view, these cells present the classical receptive field with center/surround effect. But their temporal response is different: the former responds to the stimulation later than the latter (see Figure 2.10). From a theoretical point of view, these cells are supposed to help the temporal decorrelation of the visual information (Dong and Atick, 1995).

The model of Figure 2.10 is obtained by considering that the spatiotemporal response is separable into a classical spatial response to which a temporal behavior is added. In signal processing, this temporal function is traditionally obtained by a feedback loop between two temporal low pass circuits (inset of Figure 2.10).

2.5. The Striate Cortex (V1)

In humans, the cortex surface dedicated to vision represents 15% of the total cortical surface. For monkeys it reaches 50%. The first target of the geniculate neurons is the primary visual cortex, also called striate cortex

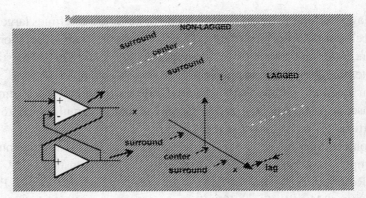

Figure 2.10. A simple model of the two kinds of cells in LGN. Responses of lagged and non-lagged cell models.

because of its appearance under the microscope, area 17 according to the classification of Brodmann, or area V1. Area V1 represents 12% of the total cortical surface of the macaque's brain, whereas it is only 3% for humans.

The visual cortex is a folded sheet of neural tissue about 2 to 3 mm thick situated in the occipital part of the brain. It is composed of six layers; the most external is layer 1; the one near the white matter is layer 6. Similar to all cortical areas, there are 3 main types of neurons: Pyramidal cells (excitatory, 70% of cells); spiny stellate cells (excitatory, 2 to 3% of cells) and inhibitory interneurons (20%). Pyramidal neurons make local connections and are the only output cells of the cortex to other cortical areas or out of the brain. Spiny stellate cells make local connections within 1 mm of their cell body. Inhibitory interneurons have an extension over a few $100\,\mu$. The resulting circuits are spatiotemporal filters that are more specific than in the retina: they select orientations, directions, spatial frequencies, colors, etc.

As the retina and the LGN appear as signal-conditioning units, area V1 can be seen as a signal analyzer. It maps the visual field according to a particular topology and decomposes each part of the retinal image into many local features.

2.5.1. *Nature and Topology of V1 Inputs*

V1 contains a retinotopic representation of the contralateral hemifield, as in LGN, but with an over representation of the central part of the retina

(Dow *et al.*, 1981):

- 80% of V1 devoted to central 10 deg
- 25% of V1 devoted to central 2.5 deg

Experimental data about the projection of the visual field on V1 have been modeled by a complex logarithm projection (Schwartz, 1980) according to the following formula:

$$z = \frac{\ln(a + \rho e^{i\theta})}{\ln(b + \rho e^{i\theta})}. \tag{2.1}$$

The real and imaginary parts of the complex number z represent the coordinates of a point on the cortical surface. The polar coordinates ρ and θ of a point on the retina give through equation (2.1) the corresponding point on the surface of V1 (Figure 2.11).

At center $\rho = 0$, the mapping is nearly linear. At larger eccentricities, radial directions on the retina are mapped onto horizontal lines on V1, and concentric circles on the retina are mapped onto vertical lines on V1.

This kind of mapping is known to have interesting properties with respect to motion. For example when moving one's head sideways or seeing a landscape through the window of a train, the central part of the projection, which corresponds to linearity, is translation motion. However, for ego-motion such as zooming or rotation (like flying in a plane), the peripheral projection deserves attention. When zooming, the radial velocity field is transformed into a horizontal translation movement, and in rotation, the

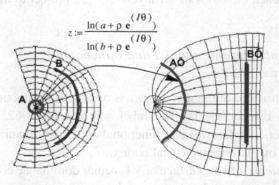

Figure 2.11. Retino-cortical projection. For a large eccentricity, radial and tangential directions respectively become horizontal and vertical.

tangential velocity field is transformed into a vertical translation movement. The general properties of the complex logarithm mapping will be studied in Chapter 6.

Usually a "cortical magnification factor" is defined as the displacement (in millimeters) on the cortical surface corresponding to 1 degree on the retina. It depends on the eccentricity r according to:

$$M = \frac{17.3}{r + 0.75}. \tag{2.2}$$

There are three parallel pathways.

The M pathway arises from α retinal ganglion cells (large neurons, large receptive field, transient, achromatic, highly sensitive to luminance contrast) and projects predominantly to layer $4C\alpha$ of V1, and weakly to layer 6.

The P pathway arises from β retinal ganglion cells (small neurons, small receptive field, rather sustained, chromatic, weakly sensitive to luminance contrast), projects predominantly to layer $4C\beta$ of V1, and weakly to layers 6 and 4A.

The K pathway arises from γ retinal ganglion cells (tiny neurons, small receptive field, rather sustained, chromatic, weakly sensitive to luminance contrast) and projects to the cytochrome oxidase blobs of layer 2/3 of V1 (Hendry and Yoshioka, 1994; Ding and Casagrande, 1997), and at least in monkey, beyond V1, to other areas such as V2, V4, MT (Bullier *et al.*, 1994). The blue-yellow ganglion cells that project to the intercalated layers of the LGN, reach the blobs of V1 layers 2/3 and project to the thin stripes of V2.

2.5.2. *Columnar Organization V1 and Functional Properties of V1*

The cortical structure of the whole brain is organized in functional columns orthogonal to the cortical layers. Hubel and Wiesel (1962, 1968, 1974) were the first ones to describe this functional columnar organization in their pioneer work on the primary visual cortex.

Hypercolumns: On the surface of V1, ocular dominance columns can be observed. They correspond to the more or less regular alternating projections of each eye. About 1 mm^2 of cortex grouping signals from both eyes form a

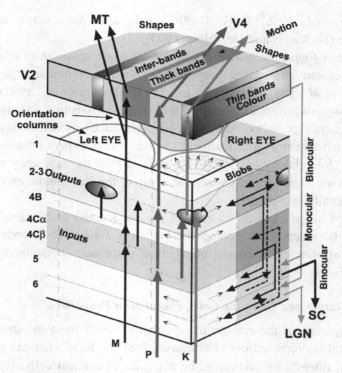

Figure 2.12. The V1 cortex and its connections from LGN and to V2. Note the columnar organization (see text). There are many intraconnections within layers and interconnections between layers (these connections make the "cortical filters").

hypercolumn. In each ocular dominance column there are cells that respond preferentially to the stimulation of one eye, but there are also "binocular cells" (Figure 2.12). Following the retinotopy, neighboring columns are related to neighboring parts of the visual field and, as the size of this part grows with retinal eccentricity, the size of the hypercolumns remains constant.

Orientation columns: Due to local circuits of connections, the cells of V1 exhibit several specific responses to the local properties of the retinal image. Hubel and Wiesel found columns of cells tuned to different stimulus orientations.

Each ocular dominance column contains 15–20 orientation columns covering the full range of orientations. In these columns, a subgroup of cells is sensitive to the direction of displacement of the stimulus

(Weliky *et al.*, 1996). Only 20–30% of the cells are strongly direction sensitive (De Valois and De Valois, 1990).

Spatial frequency columns: Columns that are sensitive to orientation are segregated into spatial frequency, again in a columnar organization (Tootell *et al.*, 1981; Bradley *et al.*, 1987; Shoham *et al.*, 1997). It has been long controversial to define the spatial disposition of orientation and frequency columns. After the early "ice cube" model, it now seems established that in higher mammal's brains, the pinwheel organization is the rule (Swindale *et al.*, 1987; Oki *et al.*, 2000): the frequency selective columns are organized as concentric coronas, the iso-orientation columns being radially dispatched.

Blobs of cytochrome oxidase are regions located in the center of the pinwheels, where cells coding for color are found. They code almost exclusively for red/green and blue/yellow chromatic oppositions and not for luminance.

2.5.3. *Simple and Complex Cells, Functional Properties*

The properties of the cells in these columns result from the structure of their local interconnections (Callaway, 1998) and form what can be called "cortical filters". By recording the activities of cortical cells, Hubel and Wiesel identified two classes of cells: the simple and the complex cells. Both classes respond to stimulations at one orientation and at one spatial frequency band.

Simple cells respond to stimuli in a linear manner and they are well represented by Gabor filters (Jones and Palmer, 1987; Heeger *et al.*, 1996; Deco and Schürmann, 2001; Ringach, 2002). Accordingly, they belong to two sub-types: in-phase cells and in-quadrature cells. Gabor filters will be studied in more detail later. For now, we will only consider their response to an impulse as a sinusoid curve weighted by a Gaussian. For in-phase cells, the curve is a cosine and for in-quadrature, it is a sine. These cells have two types of temporal behavior, their spatiotemporal response is either with separable time and space variables, or with inseparable variables (De Angelis *et al.*, 1993). The cells whose response is with inseparable variables are good candidates for motion detection.

Complex cells respond to stimuli in a non-linear manner. Their response to an adapted stimulus is the same whatever its position within the receptive field of the cell is. For simple cells, the response follows the position of the stimulus. Their response is well simulated by the sum of the squared responses of in-phase and in-quadrature simple cells.

Both simple and complex cells are suitable to code for textures: the particular distribution of activities along the cells in a hypercolumn is representative of the particular texture present at the corresponding location in the visual field.

It is well known that area V1 receives modulatory feedback from other areas. The result is that the receptive fields of cells may change dynamically according to the context (Pettet and Gilbert, 1992; Das and Gilbert, 1995). It has also been shown that, in this very first stage of cortical processing, some cells are capable of responding to complex boundaries: Stimuli like bars constituted either of simple line or of texture change result in the same responses linked to the global orientations of the stimuli (Leventhal *et al.*, 1998). This allows for the difficult task of detecting textured objects on a textured background.

There are also binocular interactions between cells in V1, which make them sensitive to depth (Poggio and Fischer, 1977). The model of Gabor filters is an interesting basis for stereo vision (Woergotter, 1999; Prince *et al.*, 2002).

2.6. Visual Areas Beyond V1 and Their Functions

Visual perception is described through a hierarchy of visual areas starting from V1 for the low level and reaching the highest levels in frontal and inferotemporal cortex (Felleman and van Essen, 1991). It has been traditionally considered that these areas were successively activated by a feedforward stream from lower to higher levels. A modern view considers that a path between two successive levels contains both feedforward and feedback connections.

2.6.1. *The Most Important Areas*

2.6.1.1. *Area V2*

Area V2 is comparable to area V1 with somewhat larger receptive fields. Using cytochrome oxidase (CO) staining, a structure of dark and pale stripes appears clearly, with specific functional properties:

- the thick dark stripes are fed by fibers issued from layer 4B of V1, linked to the magnocellular pathway.
- the thin dark stripes receive fibers from the CO blobs of V1; receptive fields are of color double-opponent type.
- the pale interstripe cells are orientation selective, they are linked to the interblobs regions, highly dominated by the parvocellular pathway.

In pale and thick stripes several authors (von der Heydt and Peterhans, 1989; Leventhal *et al.*, 1998) have found cells responding to illusory contours such as Kanizsa triangle (Kanizsa, 1979). Thick stripes are considered to be devoted to motion, thin stripes to color and interstripes to form.

It is worth noticing that V1 and V2 are strongly interconnected by reciprocal links: for example the activity of V1 can be seriously impaired if V2 is artificially inactivated (Bullier and Hupé, 1996).

2.6.1.2. *Area V3*

Originally thought to be homogeneous, this area has been recently divided into two parts: V3d (dorsal) and V3v (ventral). V3d receives inputs from V2 and directly from V1, its cells are more direction-sensitive. V3v receives inputs mostly from V2. From this point on the visual treatment separates into two fundamental streams.

2.6.1.3. *The Ventral Stream*

The ventral stream represented in light grey in Figure 2.13. V4, receives inputs from V3d and V2, as well as from the foveal region of V1. It is retinotopically organized, with receptive fields much larger than in V1 or V2. Cells in V4 are sensitive to color, form and texture of visual stimuli and give the same response whatever the displacement of the stimuli. They respond to more complex stimuli than V1 cells and are insensitive to simple stimuli that activate V1 (Gallant *et al.*, 1996). In case of multiple stimuli, the activity of neurons in V4 is modulated by attentional tasks, so that only one object is analyzed at a time (Desimone and Schein, 1987; Desimone and Duncan, 1995).

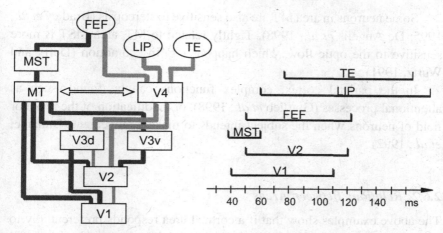

Figure 2.13. Visual cortex and the two streams. Left: relations between areas. Right: latencies of some areas after the onset of a visual stimulus, after Novak and Bullier (1997), with kind permission of Springer Science and Business Media.

The main target of V4 outputs is the inferotemporal cortex (IT), but projections to LIP and FEF are also found. The part of infero-temporal cortex fed by the parvocellular pathway is known to be specialized in color, fine patterns and fine stereopsis, IT is linked to the recognition of complex objects, neurons responding to faces have even been found (Rolls, 1992 and 1994).

2.6.1.4. *The Dorsal Stream*

The dorsal stream is represented in hatched black in Figure 2.13. The visual area V5, also called Medio-Temporal (MT) is particularly linked to the magnocellular pathway. It receives inputs from V1, V2, V3d and V3v. The neurons of MT are sensitive to the direction of motion (Albright *et al.*, 1984; Salzman *et al.*, 1992) of the objects in their receptive fields, but do not modify their response if the form or the color is changed. MT has been initially considered as the area of motion, but in monkeys the destruction of MT revealed only a temporary loss of ability to perceive directions of motion. In fact, the "center of motion" is rather a pool of neighboring areas MT-MST-FST. Similarly for V4, its destruction does not fully impair the perception of colors. These facts will be commented in the next section.

Some neurons in area MT are also sensitive to stereopsis (Bradley *et al.*, 1995; De Angelis *et al.*, 1998). Tightly related to MT, area MST is more sensitive to the optic flow, which happens e.g. in egomotion (Duffy and Wurtz, 1991).

In the parietal cortex, complex functions can be found, such as attentional processes (Gottlieb *et al.*, 1998), or modification of the receptor field of neurons when the subject intends to make a movement (Duhamel *et al.*, 1992).

2.6.2. *Relations Between Areas*

The above examples show that if a cortical area responds preferentially to some aspect of the stimulus, it shares this specificity with neighboring areas. Even areas such as MT and V4 are linked though they are specialized in very different features. Also, many areas respond to visual motion (Dupont and Orban, 1994). From another point of view, even the pair V1–V2 could be considered as a processing complex according to the findings of Bullier and Hupé (1996): inactivation of V2 severely impairs the activity of V1.

Another aspect of cortical processing is that the hierarchical organization is not respected: high-level areas are often directly targeted by first- or low-level areas (see the scheme of Figure 2.13). An important consequence is that an nth level area does not obligatorily respond after the one at level $(n - 1)$. Novak and Bullier (1997) have collected experimental data among various authors about the latency of response of neurons in different cortical areas. Later, Schmolesky *et al.* (1998) have obtained results including LGN M and P pathways. An abstract of their results is given in Figure 2.13: the dorsal stream is faster than the ventral one, and in some cases a neuron of a high-level area may respond before a neuron of a low-level area!

What is the need for such anticipation? We never question why, for example, a boss, before giving orders, needs to know exactly what kind of work he intends to give to get accomplished by his workers. It is obvious that this knowledge is essential for the work to be well accomplished. In the very same way, in cortical organization, there are as many feedback connections between areas as feedforward connections (Bullier, 2001a, 2001b).

Feedback connections are the best candidates for rapid and long-distance information exchange between neurons of distant regions in the visual field. Signals issued from the magnocellular layers of the LGN are first sent in the parietal cortex that is rapidly "aware" of the visual stimulus. The processed information is then sent "back to areas V1 and V2 that act as "active black-boards" for the rest of the visual cortical areas: information retroinjected from the parietal cortex is used to guide further processing of parvocellular and koniocellular information in the inferotemporal cortex" (Bullier, 2001 b).

Typically, feed-forward connections are responsible for the transfer of the information processed in a cortical area to higher level areas. Feedback connections play a role in the modulation of the treatments in lower-level areas, for example in modifying the extraction of a local feature according to the surrounding (contextual) information detected at a higher level, or according to some attentional process: trying to find a mushroom in the forest requires selecting the features for round and red, white or brown objects, it requires also to be less sensitive to elongated green or dark objects (grass or roots) and to oval brown objects (dead leaves). In V1 neurons Zipser *et al.* (1996) have found contextual modulation in a wide spatial extent (8° to 10° diameter), that is, far beyond the normal receptive field.

In particular, the work of Pascual-Leone and Walsh (2001) suggests that feedback connections may be essential for conscious vision. With pulses applied by transcranial magnetic stimulation (TMS) to the region of area MT in humans, it is possible to elicit the perception of moving colorless phosphenes,[8] whereas by stimulation of V1, colored static phosphenes are perceived. In order to study the role of V1 (and of the link V1-MT) in moving phosphenes, they used a double-pulse stimulation, one for V1 and one for MT.

In a first experiment, they applied the pulse on V1 before the pulse on MT, with variable delays and they asked the subjects to report their perception. Contrarily to what was expected, there were no changes of perceived phosphenes in this condition. Then they decided to stimulate V1 *after* the stimulation of MT. In this case, when V1 was stimulated

---

[8]Phosphenes are visual illusions that can be generated for example by pressing your eye: you "see" various kinds of forms in your visual field.

between 5 and 40 ms after MT, the perception of moving phosphenes was dramatically disrupted.

This experiment and others of theirs show that it is the arrival of MT descending signal to V1, which makes possible the conscious perception of phosphenes by V1 neurons. Lamme and Roelfsema (2000) had previously presented some arguments in this sense, now an experimental verification is available.

Chapter 3

Basic Model of the Retina

In this chapter we will present a simplified model of the retinal circuits. Starting from the electrical properties of the cellular membrane, we will build a linear model of the retinal cells suitable to derive the fundamental equations of the retina: those of a spatiotemporal filter.

3.1. Electrical Model of the Cellular Membrane

The cellular membrane of a cell is composed of a double sheet of lipoproteins, the thickness of which is about 70 Å (Angströms) as shown at Figure 3.1. It separates the inner and outer media, which are mainly composed of a solution of hydrated ions in water: sodium (Na^+), potassium (K^+) and chloride (Cl^-). The phospholipidic parts of both sheets face each other and are hydrophobic, they prevent the presence of water and thus the presence of ions in the membrane. This confers to the membrane its fundamental property of electrical insulation. At particular locations on the membrane there are pores that may allow some species of ions to cross the membrane under the control of various molecules.

Linked to the metabolism of the cell, an "ionic pump" controls the ionic concentration of the inner medium, mostly by insuring an outward flux for Na^+ coupled with an inward flux for K^+.

3.1.1. Electro-Osmotic Equilibrium

At rest, the membrane is semi-permeable: it may be traversed by the ionic species under the double influence of the electrical field and of the concentration gradient. For any ion, if the concentrations are $[C_1]$ in a

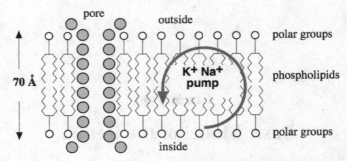

Figure 3.1. The cellular membrane as a double layer of lipoproteins separating the inner and outer media of the cell. See the text.

medium (1) and $[C_2]$ in a medium (2) separated by a semi-permeable membrane, a diffusion flux takes place from the medium of higher concentration to the one of lower concentration.

This flux is proportional to the logarithm of the concentration ratio. Because the ions are charged (either + or −), it results in a difference of electrical potential between the two media, and an electrical flux depending on the sign of the ion. In the absence of an active process, the law of Nernst states that both fluxes are equilibrated, leading to a relation between the concentration ratio and the difference of electrical potential. For the cellular membrane, this relation is:

$$E_{in} - E_{out} = -\frac{kT}{nq} \ln \frac{[C_{in}]}{[C_{out}]}, \tag{3.1}$$

k being the Boltzmann constant ($1.38 \ 10^{-23}$ J/K°), T the absolute temperature, q the elementary electrical charge ($1.6 \ 10^{-19}$ Coulomb) and n the number of charges of the ion (kT/q is about $25\,\text{mV}$ at normal room temperature and $26.7\,\text{mV}$ at mammalian body temperature).

Table 3.1 gives the Nernst potentials for the three ionic species of the cellular membrane. It can be seen that Equation (3.1) is fairly well verified for potassium and chloride, but not for sodium (opposite sign). This means that there is an active process concerning this ion. In fact, this active process for sodium is coupled to a weaker one for potassium: it is called "the Sodium-Potassium Pump". This pump is driven by the degradation of the Adenosine Tri-Phosphate (ATP) under the control of the cell's metabolism.

Table 3.1. Ionic concentrations and Nernst potentials for the giant axon of squid (according to G. Adam, 1970).

Ion	IN (mMol/l)	OUT (mMol/l)	En (mV)
Na^+	50	440	+54
K^+	400	20	−75
Cl^-	60	560	−57
Ca^{++}	0.4	10	—
Mg^{++}	10	54	—

3.1.2. *Membrane Potential, Response to a Stimulation*

We owe the first in-depth study of the membrane potential to Hodgkin and Huxley (1952). To compute the membrane potential at rest when there are several ion species in the surrounding media, it is necessary to take into account the conductances G_i and the active potentials E_i associated to each ion. In fact, the insulating property of the membrane implies the existence of a capacitance C_m which leads to a time constant. The time variation of the membrane potential $V_{i/o}(t)$, inside with respect to outside, is given by Equation 3.2:

$$C_m \frac{dV_{i/o}(t)}{dt} = [E_{Na} - V_{i/o}(t)]G_{Na} + [E_K - V_{i/o}(t)]G_K - V_{i/o}(t)G_L.$$

$$(3.2)$$

The conductance G_L stands for a leakage term due to Cl^- and other ions. When time tends to infinity, we obtain the membrane resting potential given by Equation 3.3:

$$V_{i/o} = \frac{E_{Na} G_{Na} + E_K G_K}{G_{Na} + G_K + G_L}.$$

$$(3.3)$$

In most of the living cells, this resting potential is about 60–70 mV, negative inside with respect to outside.

The Nernst potentials are relatively constant, however the ionic conductances are variable, linked to the aperture of the pores. In nervous cells (as in receptors), a cascade of biochemical reactions may control the aperture or the closure of pores. In chemical synapses, the presynaptic neuron releases

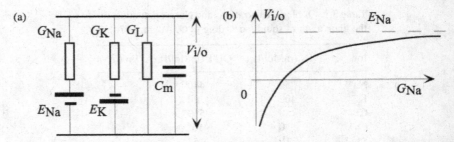

Figure 3.2. (a) Electrical model of the cell membrane; (b) Variation of the membrane potential as a function of an ionic conductance, here G_{Na}. Notice the non-linear aspect of the response.

a chemical mediator able to react with allosteric molecules tied to the pores, closing or opening them. When the result of this action drives the potential to less negative or to positive values (depolarization), the synapse is said to be "excitatory". When the potential tends to be more negative (hyperpolarization), the synapse is said to be "inhibitory". Generally, an excitatory action results from the increase of both conductances G_{Na} and G_K, whereas an inhibitory action results from the increase of G_K alone. Figure 3.2 shows an example of the membrane potential variation when G_{Na} increases.

For most of the nerve cells, if the membrane potential reaches some threshold value (about 10 mV above the resting value), the membrane turns into an active process and the potential exhibits a relaxation oscillation called the "action potential" or "spike", as shown in Figure 3.3. This very narrow pulse (0.5 ms) propagates along the axon and initiates the release of chemical mediators at the nerve ending.

In the primate's retina, this phenomenon does not appear, except for ganglion cells and some amacrine cells. The retinal neurons of the outer plexiform layer exhibit only graded potentials.

The case of photoreceptors is particular. In the dark, they are depolarized and deliver (excitatory and inhibitory) chemical mediators to the bipolar and the horizontal cells, acting as presynaptic neurons. When illuminated, the photoreceptors are hyperpolarized and they reduce their action onto their target cells. The whole process of phototransduction will be studied in detail in Chapter 6.

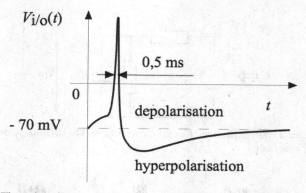

Figure 3.3. The relaxation oscillation of the action potential (spike). Once a certain threshold is exceeded, the membrane potential produces an impulse which propagates along the axon.

3.1.3. *Linear Model*

The circuit of the electrical model of the membrane (Figure 3.4) can be reduced to a unique voltage generator E_g in series with an internal resistor r_g feeding in parallel the membrane resistance r_m and the membrane capacitance C_m. The values of these equivalent generator voltage and internal resistance are given by Equation 3.4, respectively:

$$E_g = \frac{E_{Na}G_{Na} + E_K G_K}{G_{Na} + G_K} \quad \text{and} \quad r_g = \frac{1}{G_{Na} + G_K}. \tag{3.4}$$

Under normal considerations, E_g and r_g are non-linear functions of the stimulus s through the conductances $G_{Na} = G_{Na0} + \alpha s$ and $G_K = G_{K0} + \alpha' s$.

In order to simplify the computations, it will be useful to consider that E_g is proportional to s, for example $E_g = E_{g0} + \beta s$, and that r_g is constant. This would be true if we consider only small variations of s. After having used this approximation in the computations, it will be easy to re-consider the original equations and see what the non-linearity implicates in the global behavior. The equivalent circuit is given in Figure 3.4.

3.2. Gap Junctions and the Basic Retinal Circuit

We have seen in Chapter 2 that the photoreceptors, as well as the horizontal cells are linked through electrical synapses called "gap junctions". At such

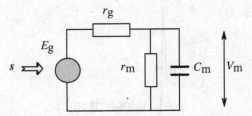

Figure 3.4. Linear equivalent circuit for the cell membrane. The stimulus s controls directly the equivalent generator.

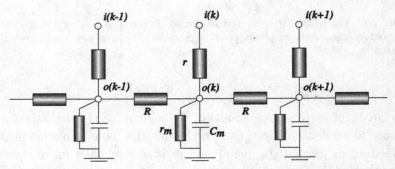

Figure 3.5. Electrical circuit equivalent to a series of cells connected by gap junctions (in one spatial dimension).

junctions, the two connected cells share the same part of the membrane so that the potential in one cell is related to the potential in the other one, through a simple resistor R as shown in Figure 3.5.

3.2.1. Basic Electrical Model and Fundamental Equation

In the outer plexiform layer of the retina, the horizontal cells are arranged in a two-dimensional array and make gap junctions with their neighbors. The same pattern of connections is supposed to be true for the photoreceptors. This kind of pattern represents the basic neural circuit of the retina, the generic electrical model of which is shown in Figure 3.5. It is in one dimension (Beaudot *et al.*, 1993).

In this Figure, a cell at position k receives a stimulus, which represents the equivalent voltage generator previously called E_g, of which the internal resistor is r, previously called r_g. The membrane resistor and capacitor are

r_m and C_m respectively. The gap junctions between cells are represented by resistors R. The (output) membrane potential of each cell at position k is $o(k)$. The set of signals $\{i(k), i(k+1)\ldots\}$ will be called the input signal of the circuit. In fact, we must consider that it is time-dependent and write it as $\{i(k,t), i(k+1,t)\ldots\}$. Similarly, the output signal (membrane potentials) will be noted $\{o(k,t), o(k+1,t)\ldots\}$.

Both input and output signals are said to be spatiotemporal. They depend on the discrete spatial variable k and on the continuous variable t. In a first step, we will consider that the cells are equidistant, situated at every $k\Delta l$. To simplify, we will take this interval as unitary: $\Delta l = 1$.

In order to calculate the expression of the output signal with respect to the input signal, let us write that the sum of the incoming currents at each node is null:

$$\frac{i(k,t) - o(k,t)}{r} + \frac{o(k-1,t) - o(k,t)}{R} + \frac{o(k+1,t) - o(k,t)}{R}$$
$$- \frac{o(k,t)}{r_m} - C_m \frac{do(k,t)}{dt} = 0. \tag{3.5}$$

3.2.2. *Spatiotemporal Response*

3.2.2.1. *Frequency Representation*

The direct general solution of this equation is not simple. A good mean is to replace each term by its Fourier transform. The double Fourier transform of the input signal $i(k,t)$ with respect to the discrete variable k and the continuous one t is: $I(f_s, f_t)$, similarly, the double Fourier transform of the output signal $o(k,t)$ is $O(f_s, f_t)$. The new variables f_s and f_t are respectively the spatial frequency in cycles per unit of length (Δl) and the temporal frequency in cycles per unit of time (measured in seconds). Due to the properties of the Fourier transform with respect to continuous time derivation $d(\cdot)/dt$ and to spatial shift $k \rightarrow k-1$, the new expression in the Fourier domain for Equation 3.5 is:

$$\frac{I(f_s, f_t) - O(f_s, f_t)}{r} + \frac{e^{-j2\pi f_s} O(f_s, f_t) - O(f_s, f_t)}{R}$$
$$+ \frac{e^{+j2\pi f_s} O(f_s, f_t) - O(f_s, f_t)}{R} - \frac{O(f_s, f_t)}{r_m}$$
$$- j2\pi f_s C_m O(f_s, f_t) = 0,$$

and, with suitable factorizations:

$$O(f_s, f_t) = \frac{I(f_s, f_t)}{1 + \frac{r}{r_m} + 2\frac{r}{R} - \frac{r}{R}(e^{+j2\pi f_s \Delta l} + e^{-j2\pi f_s \Delta l}) + j2\pi f_s r C_m f_t}.$$

By posing $\alpha = r/R$ $\beta = r/r_m$ and $\tau = rC_m$, and applying Euler's formulae, we obtain:

$$G(f_s, f_t) = \frac{O(f_s, f_t)}{I(f_s, f_t)} = \frac{1}{1 + \beta + 2\alpha(1 - \cos(2\pi f_s \Delta l)) + j2\pi \tau f_t}.$$
$$(3.6)$$

This ratio $G(f_s, f_t)$ between the Fourier transforms of the output and input signals is called the spatiotemporal "Transfer Function" of the circuit. It is useful in order to characterize what kind of processing the input signal undergoes when passing through the circuit.

This processing is well represented by a graphical drawing of the two-dimensional function $G(f_s, f_t)$, as shown in Figure 3.5.a: It is clear that this function is maximum at zero frequencies and then decreases to zero for high frequencies. For this reason, this circuit is called a "spatiotemporal low-pass filter". This means that the faster a signal varies (spatially or temporally), the more it appears attenuated at the output.

Let us note that this function is periodic in spatial frequency (Equation 3.6): this property is due to the discrete nature of the spatial variable. The range of spatial frequencies to be considered is between $-1/2$ and $1/2$ of the spatial sampling frequency $(1/\Delta l)$. For humans, the spatial sampling frequency $1/\Delta l$ corresponds to 60 cycles per degree.

3.2.2.2. *Spatial and Temporal Behavior*

The inverse Fourier transform of the transfer function gives the impulse response of the circuit, that is, the values taken by when the input is a Dirac impulse in time applied at the single location $k = 0$. The following formula gives an approximation of the impulse response of our basic circuit:

$$g(k, t) = \frac{1}{2}\left(\pi \frac{\alpha \Delta l^2}{\tau} t\right)^{-\frac{1}{2}} e^{-\frac{t}{\tau}} e^{-\frac{k^2}{4\frac{\alpha \Delta l^2}{\tau} t}}.$$
$$(3.7)$$

This result suggests a diffusion phenomenon with loss. For a diffusion process without loss, the spatiotemporal variation of concentration $c(x, t)$

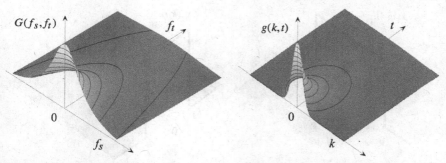

Figure 3.6. Spatiotemporal behavior of the basic circuit. (a) Fourier domain: the low-pass spatiotemporal transfer function; (b) Signal domain: the spatiotemporal impulse response. Note the effect of diffusion.

in a diffusion medium is:

$$c(x, t) = \frac{1}{\sqrt{\pi D t}} e^{-\frac{x^2}{4Dt}},$$

D being the diffusion constant in the medium. This result shows that the potential in one cell at a given position will diffuse in the neighboring cells, while vanishing with time. In other words, it means that the output signal will follow the input signal, but with some temporal lag and with a spatial smoothing of the input, as shown in Figure 3.6.

3.2.2.3. *Spatial and Temporal Inseparability*

Let us observe a very important property of this model: the spatial and temporal inseparability. Looking at both the formulae of the transfer function (Equation 3.6) and that of the impulse response (Equation 3.7), it is clear that temporal and spatial variables do not appear alone in a single term: the response is not the product of a spatial term by a temporal term.

In other words, the form of the spatial response depends on the temporal content of the input signal, and *vice versa*. The form of the temporal response depends on the spatial content of the input signal (Figure 3.7).

As a consequence of this inseparability, the behavior of this circuit will be of special interest:

- The spatial filtering is not static: it will evolve with time after a stimulation has been applied.

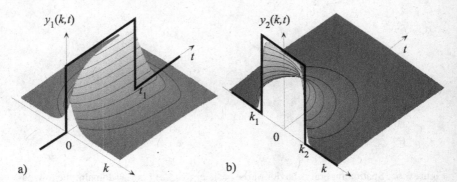

Figure 3.7. Spatiotemporal response $y(k, t)$ when the input signal $x(k, t)$ is: (a) a rectangular function in time between $t = 0$ and $t = t_1$ applied at a unique location $k = 0$; (b) a rectangular function in space between locations k_1 and k_2 and a Dirac function in time at $t = 0$. Observe in both cases the spatial smoothing and the temporal lag which are typical of low-pass filters.

- When stimulated by an input signal where time and space are coupled, for example a moving stimulus $x(k, t) = x(k - vt)$ of velocity v, the response will be a motion signal of the variable $k - vt$ with the same velocity. Its spatial form is that of the input filtered by replacing the term in t by k/v in Equation (3.7), or the term in f_t by $v f_s$ in Equation (3.6).

We will see later (Sections 3.3.1.4 and 3.5.2) that this property is of crucial importance. It confers to the retina its very specific spatiotemporal behavior, which is at the origin of a "coarse-to-fine" process, together with a special processing of motion in the retinal image.

3.3. Models of the Outer Plexiform Layer

In Chapter 2, Section 2.3.2, we have seen the fundamental structure of the OPL (Outer Plexiform Layer): the synaptic triad. First, the incoming light activates the photoreceptor (we will say "activate", even if the action of light is a depolarization). Then, the photoreceptor excites the horizontal cell and the ON bipolar cell, and inhibits the OFF bipolar cell (Dacey *et al.*, 2000).

According to the biological data, there are two possible structures of connections (see Chapter 1). In the *feed-forward* model (Yang *et al.*, 1991), the horizontal cell inhibits the ON bipolar cell and excites the OFF bipolar cell. Because of the symmetry between the ON and OFF bipolar cells, we will consider a theoretical model where there is only one kind of bipolar cell. In this model, the signal is the difference between the ON and OFF signals. In the *feed-back* model (Verweij *et al.*, 1996), the horizontal cells return an inhibiting signal back to the photoreceptors.

We will study the possibility of a third structure: a compound model of the two previous structures, which exhibits interesting filtering properties.

3.3.1. *The Basic Feed-forward Model*

3.3.1.1. *Equation*

As we have said, both photoreceptors and horizontal cells respond to the basic gap-junction sub-circuit. The transfer functions of these sub-circuits are described by Equation 3.8, with subscripts "*r*" and "*h*" for photoreceptors and horizontal cells respectively:

Because the two kinds of cells have different geometrical sizes, they may have different coefficient values for spatial (α, β) and temporal (τ) constants.

$$G_r(f_s, f_t) = \frac{1}{1 + \beta_r + 2\alpha_r(1 - \cos(2\pi f_s \Delta l)) + j2\pi \tau_r f_t}$$

$$G_h(f_s, f_t) = \frac{1}{1 + \beta_h + 2\alpha_h(1 - \cos(2\pi f_s \Delta l)) + j2\pi \tau_h f_t} \tag{3.8}$$

For the feed-forward model (according to the scheme of Figure 3.8(a)), and in the Fourier domain, the outputs $Y_b(f_s, f_t)$ of the bipolar cells are expressed in function of the inputs $X(f_s, f_t)$ and of the transfer functions $G_r(f_s, f_t)$ and $G_h(f_s, f_t)$:

$$Y_b(f_s, f_t) = X(f_s, f_t)G_r(f_s, f_t) - X(f_s, f_t)G_r(f_s, f_t)G_h(f_s, f_t).$$

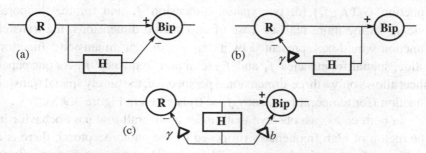

Figure 3.8. The two structural models: (a) Feed-forward; (b) Feed-back; (c) Compound feed-forward/feed-back.

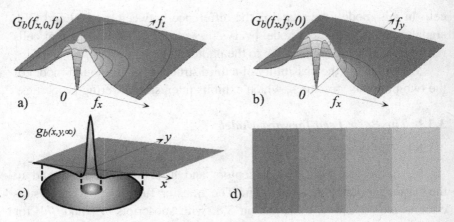

Figure 3.9. The high-pass behavior of the feed-forward model in (a) the spatio-temporal frequency domain and in (b) the spatial frequency domain. In (c) the spatial impulse response (for $t = \infty$) is shown, i.e. the receptor field of a bipolar cell. In (d) the phenomenon of Mach bands.

Then, the transfer function at the output of the bipolar cells takes the following form.

$$\frac{Y_b(f_s, f_t)}{X(f_s, f_t)} = G_b(f_s, f_t) = G_r(f_s, f_t)[1 - G_h(f_s, f_t)]. \quad (3.9)$$

3.3.1.2. *Spatiotemporal Transfer Function of the Feed-Forward Model*

The behavior of this circuit will be studied first in the Fourier domain. The first term of Equation (3.9) gives an overall low-pass behavior. The second term, subtracting a low-pass version of its input, will provide a further high-pass behavior. The drawing of Figure 3.9(a) shows the resulting transfer function $G_b(f_s, f_t)$ for one spatial dimension f_x and for the temporal one f_t. In the more realistic case of two spatial dimensions, the transfer function would be represented by a drawing in four dimensions: function value, spatial frequencies f_x and f_y, temporal frequency f_t. As our paper sheet allows only a three-dimensional perspective, the purely spatial transfer function (for temporal frequency $f_t = 0$) is given in Figure 3.9(b).

In both cases, we observe that, under an overall low-pass behavior in the region of high frequencies (imposed by the photoreceptors), there is a high-pass function in the region of low spatial and temporal frequencies (due to the inhibitory action of the horizontal cells). The global shape of this transfer function in the spatial frequencies, as in the spatiotemporal frequencies is that of a volcano. Practically:

- the high spatial frequencies lie from 1/4 to 1/2 of the sampling frequency (around 30–60 cycles per degree in human), that is roughly the frequencies that are concerned when you look at this text from a distance of 2 m.
- the high temporal frequencies start from about 10 Hz and reach the limit of flicker fusion around 20 Hz.

At this stage, two questions may arise about the retinal filter:

- *Why are the photoreceptors coupled in a low-pass filter?*

The coupling of photoreceptors may seem to be a bad solution for image processing: the low-pass resulting filter will add some blur to the retinal image, thus degrading it before further processing! This is contrary to the idea of extracting as much information as possible from the visual world.

In fact, let us remember that it is impossible for the photoreceptors to be accurately identical to each other: each cell has its own individual metabolism with its own geometry slightly different from those of its neighbors. This leads to individual behaviors with different thresholds and gains among the receptors in the conversion of light into electrical signals. Thus the model to be considered is a set of ideally identical receptors receiving the input image and applying to it additive and multiplicative noises, representing the mismatch between individual thresholds and gains. Because this noise changes independently from cell to cell, it represents a random signal in the range of very high spatial frequencies.

The coupling of photoreceptors creates a spatial low-pass filter capable of smoothing this added noise around the spatial sampling frequency. It must be said that the useful range of frequencies for a visual scene is far smaller than the sampling frequency, and that it is weakly affected by this smoothing.

It has also be observed that this filter is not spatially constant: the coupling resistors of gap junctions may also be mismatched. In his Ph.D. thesis, Torralba (1999) has shown that the spatial filters built from a network of resistors are very robust to individual dispersion of values: a random mismatch of ±20% around the mean resistance value leads to an output noise of maximum ±2% with respect to the ideal filter response.

- *Why does the retina behave as a high-pass filter?*

If you ask somebody to draw a sheep, you will more certainly get the outline of the animal rather than a graded picture with different shades. This means that the contours of objects are more important for recognition than their shaded surfaces.

The main function of spatial high-pass filtering is precisely to enhance contours. Another fact to consider is that objects' surfaces may be submitted to a large range of illumination, widely changing their luminance distribution and appearance, whereas the contours remain more unchanged. A well known effect of this spatial high-pass filtering, is the "Mach bands phenomenon": when looking at the boarder between two levels of gray, we observe two transition zones on each surface near the boarder, one appearing brighter on the side of the brighter surface, and the other one appearing darker on the side of the darker surface (Figure 3.9(d)). Near boarders, the perceived contrast is stronger than the effective contrast.

Besides, the statistics of natural images tells that their mean frequency spectrum is shaped in $1/f$, either for the spatial frequencies in static images or for the spatiotemporal frequencies in moving sequences (Atick and Redlich, 1992). This means that the lower frequencies carry a very strong power that may cancel the perception of higher frequencies. The role of the retinal high-pass filter is to compensate this $1/f$ spectrum. In signal processing, this technique is called "spectral whitening".

3.3.1.3. *Comparison with Biological Data*

The biological literature provides a large amount of measurements of the visual spatial and temporal transfer functions (Buser and Imbert, 1992). For the retinal response, experiments are carried on animals. For humans, only the global response is available, through psychophysical experiments.

Figure 3.10 gives the fit of the model to experimental data, in both cases, the model is provided with the same parameters. For the spatial transfer function, the three degrees of freedom of the model are sufficient (β_r has no importance): α_r gives the curve in the high frequencies, α_h gives the high cut-off frequency (around 2 $c/d°$) and β_h with α_h gives the low cut

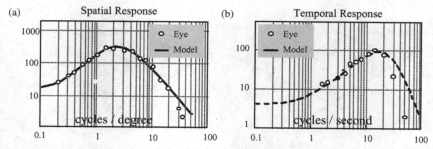

Figure 3.10. Comparison between the model (black line) and biological data. (circles). (a) spatial transfer function; (b) temporal transfer function.

off frequency. For the temporal transfer function the model's time constants can explain the rising part of the curve, but additional time constants are necessary to explain the high frequency decrease. This is not surprising for these data are derived from perception experiments, hence they involve more than the retina and the additional time constants may be due to other processing layers like LGN and cortical structures.

3.3.1.4. *Consequence of Time-Space Inseparability*

As we have already mentioned, the inseparability of temporal and spatial variables in the transfer function implies that the spatial behavior will depend on the temporal content of the input signal (and *vice-versa*).

In order to illustrate this property, let us imagine that the input signal is an image in front of the eye, but in the dark. Suddenly the light is turned on at time $t = 0$ and turned off a while after. By taking the inverse Fourier transform only with respect to temporal frequency, we obtain the temporal evolution of the spatial transfer function when the temporal aspect of the stimulus is a square wave. Figure 3.11(a) gives, for one spatial dimension the time-course of the bipolar cells' transfer function.

It is interesting to observe that just after the light is turned on, the spatial behavior of the OPL is low-pass. It progressively becomes high-pass as time progresses. When light is turned off, a negative rebound appears only for the very low spatial frequencies. This behavior explains why the names "ON" and "OFF" are given to the two kinds of bipolar cells. Experimentally, the same behavior has been recorded in the Lateral Geniculate Nucleus (LGN) for cells belonging to the "Parvocellular pathway". In this pathway, the cells

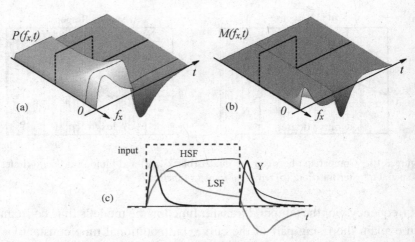

Figure 3.11. Temporal evolution of the spatial transfer function of the retina. (a) The output of bipolar cells, at the origin of the Parvocellular pathway; (b) its spatially low-pass filtered version, at the origin of the Magnocellular pathway. The duration of the stimulation is represented by a square wave (the black line); (c) The temporal profiles of high spatial frequencies (HSF), low spatial frequencies (LSF) and Y cells.

are known to respond to image details in a temporally sustained manner. We explain this property by looking at Figure 3.11(a): the time course of the high spatial frequencies (carrying details) follows the temporal input in a sustained manner, with some lag (temporal low-pass behavior). They are issued from the midget ganglion cells whose individuals connect only one bipolar cell.

In the LGN, there are other cells belonging to the so-called "Magnocellular pathway". These cells are known to behave differently compared to the previous cells. They are temporally transient and spatially low-pass. They are issued from the parasol ganglion cells whose individuals connect to several bipolar cells. This multiple connection can be seen as a purely spatial low-pass filter. In our model, by spatially low-pass filtering the bipolar outputs, we obtain a response shown in Figure 3.11(b), which presents all the characteristics of the magnocellular pathway.

It is interesting to observe that, with such a simple retinal model using a simple scheme of connections, we are able to provide biologically plausible hypotheses on the functions of the two major pathways to the visual cortical areas. We will see more details about these functions in Section 3.4.

3.3.2. *The Feedback Model*

Many biologists have explained the high-pass behavior of the OPL by means of a negative feed-back from the horizontal cells to the photoreceptors. Some of them think that it would be the essential mechanism for this behavior. We will use our model in order to verify this hypothesis and will consider now the diagram of Figure 3.8(b). The corresponding circuit can be directly translated into the following equation relating the response $Y_b(f_s, f_t)$ to the input $X(f_s, f_t)$:

$$Y_b(f_s, f_t) = G_r(f_s, f_t)[X(f_s, f_t) - \gamma G_h(f_s, f_t)Y(f_s, f_t)], \qquad (3.10)$$

γ being the (positive) coefficient of negative feed-back.

By re-arranging Equation (3.9), we obtain the new transfer function (Equation 3.10):

$$\frac{Y_b(f_s, f_t)}{X(f_s, f_t)} = G_b(f_s, f_t) = \frac{G_r(f_s, f_t)}{1 + \gamma G_r(f_s, f_t)G_h(f_s, f_t)}. \qquad (3.11)$$

The representation of this transfer function is given at Figure 3.11(a) for one spatial dimension.

In this case, in order to obtain the same amount of low frequency rejection as in the feed-forward model, we are obliged to choose a sufficiently high value for the inhibition coefficient γ, leading to a rather high peaking on the temporal frequency axis. The reason is that when a temporal low-pass system (here: receptors) is looped with another temporal low-pass system (here: horizontal cells), the resulting system is said to be of second order. Such a system may present a ringing behavior if the coefficient γ exceeds a given value. This is what happens here. The ringing behavior is illustrated by the temporal impulse response as shown at Figure 3.12(b).

In biological data, we observe some negative rebound in the temporal response of the photoreceptors, but no ringing phenomenon. This means that the feed-back model of horizontal cells certainly exists, but because there is no significant ringing, it cannot explain the low frequency rejection observed at the bipolar level.

In fact, as mentioned in Chapter 2, it is worth merging the two schools of biologists and testing a new working hypothesis, that of a compound model including both interactions from the layer of horizontal cells: feed-forward to the bipolar ones and feed-back to the photoreceptors.

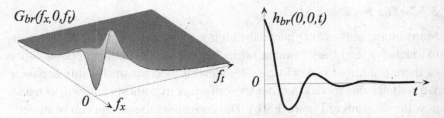

Figure 3.12. The Horizontal-to-Receptor feed-back model. (a) Spatiotemporal transfer function; (b) temporal impulse response at the output of a photoreceptor. Note the important oscillations induced by a feed-back strong enough to obtain the retinal low frequency rejection.

3.3.3. The Feed-Forward — Feed-Back Model

This model corresponds to the schema of Figure 3.8(c), where the horizontal cells, in addition to the feed-back loop, send an inhibitory connection to the bipolar cells through a (positive) coefficient b. In this case, the new transfer function for the OPL is given by Equation (3.12):

$$G_b(f_s, f_t) = G_r(f_s, f_t) \frac{1 - bG_h(f_s, f_t)}{1 + \gamma\, G_r(f_s, f_t)G_h(f_s, f_t)}. \qquad (3.12)$$

In this case it is always possible to choose coefficients γ and b in order to obtain the desired spatiotemporal transfer function without a prohibitive ringing phenomenon, so it is not necessary to illustrate it by a perspective Figure. More interesting is the spatial aspect of this transfer function. Figure 3.13 shows a comparative study of the three models of horizontal cells connectivity (feed-forward, feed-back and compound). The three curves of spatial transfer functions are scaled to have the same maximum. We can see how the compound model allows the widest range of transition from the region of low frequencies to the region of frequency of the maximum transfer. This is of utmost importance if we consider that one of the roles of the retina is to compensate for the $1/f$ spectrum of natural images (see Section 3.3.1.2).

3.3.4. The Midget and Diffuse Bipolar Model

The midget bipolar cells are connected to only one photoreceptor, consequently they respond to the above-described mathematical model. The diffuse bipolar cells are known to connect several photoreceptors, the

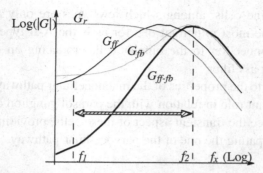

Figure 3.13. Spatial transfer functions for the three models G_{ff}, G_{fb} and G_{ff-fb}: respectively feed-back, feed-forward, and compound. G_r is the photoreceptor's transfer function. All transfer functions are normalized to their maximum. G_{ff-fb} allows the maximum transition bandwidth (double arrow) for the high-pass filtering.

number of connections depending on the eccentricity of their position. In the simplest way, they can be modeled by a purely spatial low-pass filter with a space constant linked to the eccentricity, in cascade with a purely temporal filter linked to the cell's membrane time constant. This model will not be used here because we are more concerned with the foveal part of the retina.

3.4. Model of the Inner Plexiform Layer

3.4.1. *Midget and Parasol Ganglion Cells*

In relation to the connectivity of midget and diffuse bipolar cells, we consider now the midget and parasol ganglion cells. The midget ganglion cells are connected to only one bipolar cell and thereby present exactly the same transfer function as the midget bipolar cells. Their outputs correspond exactly to the above-defined model.

The parasol ganglion cells connect several bipolar cells (depending on retinal eccentricity). Due to this connection scheme, they will be represented by a purely spatial low-pass filter in series with a purely low-pass temporal filter (membrane time constant).

3.4.2. *The Amacrine Cells*

In the inner plexiform layer (IPL), there are two sub-layers of connections, one for the ON-bipolar cells, and one for the OFF bipolar cells. In each layer we find amacrine cells, among which few types are only approximately understood. The most important one for us is the A-II type, which sends a feed-back connection to the bipolar cell, producing an overall purely temporal high-pass filter.

According to the properties of the magnocellular pathway, they seem to play an important role in relation with the parasol ganglion cells. They are likely to enhance the transient aspect of these cells, providing them with a response anticipating the one of the parvocellular pathway.

3.4.3. *ON, OFF and Bi-Stratified Ganglion Cells*

The midget and parasol ganglion cells are connected to the bipolar and amacrine cells in the sub-layer of the IPL corresponding to their ON or OFF characteristic. These cells are known to present a rather linear response (the response to the sum of two signals is the sum of the responses to each signal presented individually) they are called "X type", whatever their midget or parasol qualification.

There is another type of ganglion cells called "bi-stratified" because they make connections in both sub-layers. These cells exhibit a non-linear response and, in particular, they respond as well to the onset of light as to the offset. In cats, they are called ON-OFF cells or Y cells. They are also of the parasol type. Because of their double connection, they respond to the ON bipolar cells as well as to the OFF ones. The mathematical model of this response is that of an absolute value as explained below.

Let us call R the output of the ideal bipolar cells of our OPL model, which can take positive or negative values. The response of midget or parasol ganglion cells of ON type will be expressed by $R_{ON} = \max\{0, R\}$, that is, the positive part of R. Similarly, the response of midget or parasol ganglion cells of OFF type will be expressed by $R_{OFF} = -\min\{0, R\}$, that is, the opposite of the negative part of R, with the obvious relation: $R = R_{ON} - R_{OFF}$.

For the bi-stratified or Y or ON-OFF ganglion cells, the response will be $R_Y = R_{ON} + R_{OFF} = \text{abs}(R)$, which is typically non-linear. These cells are mainly of the parasol type, that is with a marked low-pass spatial behavior. They belong to the magnocellular pathway. We will see later that the role of the Y ganglion cells is of major importance in signaling relatively large objects or objects which suddenly appear. They project to the V1 area, but

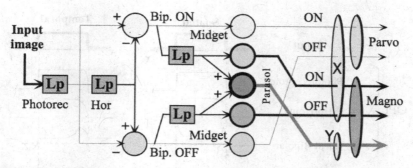

Figure 3.14. Architecture of the global model of the retina showing the circuits of OPL and IPL and the output signals.

also to the Superior Colliculus and to archaic visual structures. They may be involved in automatic tasks as gaze control or obstacle avoidance.

3.4.4. Summary of the Retinal Circuits and Output Signals

3.4.4.1. *Summary of the Retinal Architecture*

Figure 3.14 shows a symbolic diagram of the retinal architecture. Starting from the left we find first the photoreceptor layer with its low-pass filter (Lp) due to spatial coupling. Then we have the horizontal cell layer with also a low-pass filter (Lp) and the two bipolar cells types ON (top) and OFF (bottom). This constitutes the model of the Outer Plexiform Layer. In the Inner Plexiform Layer we find the ON and OFF Midget ganglion cells: they receive directly the signals of bipolar cells, they are of X type (linear behavior) and their axons constitute the parvocellular pathway. Because these tracks vehicle high-spatial frequency signals (hence details), the are drawn in thin lines.

In this IPL we find also the parasol ganglion cells, which are symbolically represented by taking a spatially low-pass version of the Midget bipolar signals. Both types of ON and OFF cells are also present; their axons constitute the X part of the magnocellular pathway. The Y ganglion cells (non-linear behavior) are represented by summing the two low-pass filters of midget ON and OFF bipolar cells. They also belong to the magnocellular pathway. We will show in Chapter 5 that the midget cells code color information in the range of very high spatial frequencies. This will explain

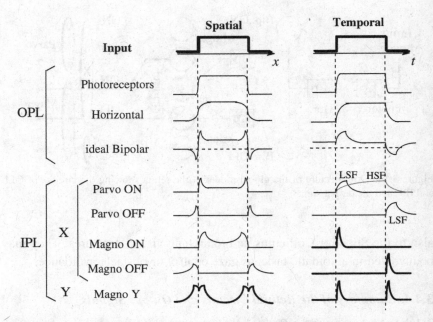

Figure 3.15. Synthesis of the retinal signals in the spatial and temporal domains. OPL: in the Outer Plexiform Layer. IPL: Inner Plexiform layer and output signals (see text).

why the cells of the parvocellular pathway respond to color whereas those of the magnocellular pathway do not (their low spatial frequency range prevents them to carry color).

3.4.4.2. *Spatial and Temporal Signals*

In Figure 3.15 we find a summary of the spatial and temporal aspects of the signals in each category of retinal cells. The input stimulus is either a spatial square wave between two spatial locations x_1 and x_2, or a temporal square wave between time instants t_1 and t_2. First of all we should note the contrast between the symmetry of the spatial responses and the asymmetry of the temporal ones. This is because the temporal response cannot precede the input variations (principle of causality), whereas a spatial input may affect the system in both directions (left or right). According to the connection scheme, the spatial response may graphically appear either symmetrical or "causal" or even "anti-causal".

Some observations:

- Because the parvocellular cells are more of high spatial frequency type than the magnocellular ones, their spatial extent is narrower.
- Because the parvocellular cells belong more to the low temporal frequency type than the magnocellular ones, their temporal lag is longer.
- In the case of parvocellular cells, the temporal response differs according to the concerned range of spatial frequencies: low spatial frequencies (LSF) appear temporally as being transient, whereas high spatial frequencies (HSF) appear temporally as being sustained.
- For Y ganglion cells, the non-linear behavior (equivalent to the absolute value) is clear.
- Due to the temporal high-pass filter added by the amacrine cells of the IPL, there is a hierarchy between the times of arrival of these signals to the cortical area: the magnocellular system reaches the cortex in advance with respect to the parvocellular one. The first wave is the one of Y cells signal. It seems to warn the brain when (and where) something is happening in the visual field. The second wave is the X-magnocellular signal that gives a rough idea of the scene components (i.e. a smoothed image where only the blobs of activity are present). And the third wave is the one of the parvocellular signal that brings information about the details of the picture and also about color. In short, the three phases are: warning information, coarse information, then detailed information. The second and third phases have been defined as a "coarse-to-fine" process.

Now we should make a parallel between the retinal processing and the classical technique of pre-processing in the domain of data analysis. In data analysis, in order to be able to compare different sets of data, first it is necessary to bring them to a same base-line: for each set, we subtract its mean value. Hence, all the data sets will show variations around a zero base-line. This procedure is called "data centering". Because the maximal amplitudes of variations may be strongly different from one set to another, the second operation to process is normalization. For this reason we compute the mean quadratic value (the variance) of each centered data and divide the data set by the obtained value. This operation is called "data reduction".

In the retina, the smoothing of the input image by the horizontal cells may be seen as operating a "local" mean. The difference between the input

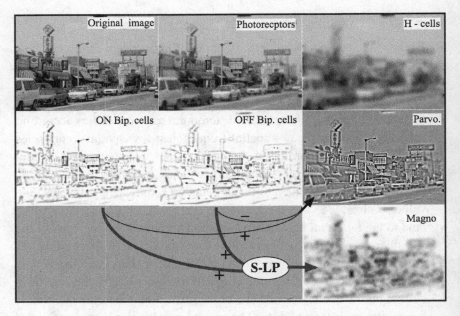

Figure 3.16.   Illustration of the spatial signals processed in the retina.

and its smoothed version appears then as a spatial "local centering" of the input image. As the operation of the Y ganglion cells is to take the absolute value and make a smoothing low-pass filtering, it can be seen as computing some equivalent of the "local variance" of the image. We do not know if the parvocellular signal is divided by the Y cells' signal to make the equivalent of a local reduction. But the idea of local variance is of some interest: the local variance is equivalent to the local energy of the signal, i.e. where there are big objects or objects with many details. Thus the Y ganglion cells are likely to tell the brain where energetic (i.e. interesting) signals are or happen in the visual field.

3.5. Response of the Model to Moving Objects

3.5.1. *Basic Formulae for Motion*

In order to understand what happens in the presence of motion, it is necessary to derive a frequency model of a moving object in space and a frequency model of the retinal filter's response.

Let the input be an image $i_s(x, y)$ with its spectral equivalent $I_s(f_x, f_y)$:

$$i_s(x, y) \xrightarrow{FT_{x,y}} I_s(f_x, f_y).$$

When a translation motion with a velocity vector $\mathbf{v} = (v_x, v_y)$ is applied to this image, the spatiotemporal input signal becomes (see Chapter 4):

$$i_s(x - v_x t, y - v_y t) \xrightarrow{FT_{x,y,t}} I_s(f_x, f_y)\partial(f_t + v_x f_x + v_y f_y).$$

Let be $g(x, y, t)$ the retinal impulse response and $G(f_x, f_y, f_t)$, the corresponding transfer function:

$$g(x, y, t) \xrightarrow{FT_{x,y,t}} G(f_x, f_y, f_t).$$

Then, the frequency response is expressed as the product of the transfer function by the Fourier transform of the input signal:

$$R(f_x, f_y, f_t) = G(f_x, f_y, f_t)I_s(f_x, f_y)\partial(f_t + v_x f_x + v_y f_y).$$

Let us take first the inverse Fourier transform with respect to the temporal frequency. We obtain the temporal evolution of the spatial frequency spectrum of the response. Because of the Dirac term, the temporal frequency is replaced by the opposite of the dot product between the velocity vector and the spatial frequency vector:

$$R(f_x, f_y, t) = e^{-j2\pi(v_x f_x + v_y f_y)t} G(f_x, f_y, -v_x f_x - v_y f_y)I_s(f_x, f_y).$$

Then let us take the spatial inverse Fourier transform to obtain the spatiotemporal response.

$$r(x, y, t) = [\partial(x - v_x t)\partial(y - v_y t)] \underset{x,y}{*} g'(x, y; v_x, v_y) \underset{x,y}{*} i_s(x, y),$$

or:

$$r(x, y, t) = g'(x - v_x t, y - v_y t; v_x, v_y) \underset{x,y}{*} i_s(x, y). \qquad (3.13)$$

The result (Equation 3.13) is a signal which moves (due to the convolution by $\partial(x - v_x t)\partial(y - v_y t)$) at the same velocity and in the same direction as the input. This signal is the spatial response of a new filter to the static image $i_s(x, y)$. This new filter is derived from the retinal filter, but its behavior is strongly dependent on the motion parameters.

Notice that the new filter is very different from the original one: $g'(x, y; v_x, v_y) \neq g(x, y, 0)$. In addition to its former spatial component, its (spatially transformed) temporal component provides some causality or anti-causality according to the direction of motion. Moreover, let's remember that the spatial and temporal variables are not separable. An example of application is given in the next section.

3.5.2. Retinal Filter with Motion

The retinal filter is represented in a simplified manner by the cascade of three terms corresponding to the OPL (triadic synapse) for the first term, and to the IPL (amacrine cells) for the second one. The third term may be due either to the OPL or to the IPL.

For the OPL, we will make the following simplifications: the photoreceptors transfer function is supposed to be unitary and the model of the OPL is feed-forward. We suppose however that the inhibition of the horizontal cells may not be fully efficient (coefficient $b \leq 1$):

$$H_{opl}(f_x, f_y, f_t) = \left(1 - b\frac{1}{1 + 4\pi^2\alpha_h(f_x^2 + f_y^2) + j2\pi\tau f_t}\right).$$

For the amacrine cells, we use the property of a high-pass purely temporal filter:

$$H_a(f_t) = \frac{1 + j2\pi\tau_1 f_t}{1 + j2\pi\tau_2 f_t}.$$

The third term is a purely spatial low-pass filter, due either to the diffuse bipolar cells or to the small parasol ganglion cells:

$$H_g(f_x, f_y) = \frac{1}{1 + 4\pi^2\alpha_g(f_x^2 + f_y^2)}.$$

In the presence of motion, the first two elementary transfer functions will be changed as previously shown, for the (first) spatiotemporal one:

$$H_{opl} \rightarrow H_{opl}^v(f_x, f_y)$$

$$= \left(1 - b\frac{1}{1 + 4\pi^2\alpha_h(f_x^2 + f_y^2) - j2\pi\tau(v_x f_x + v_y f_y)}\right),$$

and for the (second) purely temporal one:

$$H_a(f_t) \rightarrow H_a^v(f_x, f_y) = \frac{1 - j2\pi \tau_1(v_x f_x + v_y f_y)}{1 - j2\pi \tau_2(v_x f_x + v_y f_y)}.$$

The third transfer function is unchanged because it is purely spatial. The spatiotemporal response of our model is:

$$R(f_x, f_y, f_t) = H_{opl}^v(f_x, f_y) \cdot H_a^v(f_x, f_y) \cdot H_g(f_x, f_y)$$
$$\cdot I_s(f_x, f_y) \cdot \partial(f_t + v_x f_x + v_y f_y).$$

By taking the inverse Fourier transform of $R(f_x, f_y, f_t)$, we obtain the retinal response $r(x, y, t)$. It is the translation (at the same velocity and in the same direction as the input) of a spatial filtering of the static input image. In order to make the calculations more feasible, we will carry the simulation in one spatial dimension. The result is illustrated in Figure 3.17. The spatial profile of the response at a given time instant, it is compared with a biological experiment.

In Figure 3.17(a), the stimulus is a white bar of light which moves at a constant velocity in front of a black background. The response of the model is shown in Figure 3.17(b) (top: the bar is moving to the left, bottom the bar is static). In biology it is not possible to record simultaneously a large number of neighboring cells. The corresponding experiment consists in the recording of one single cell: because of the constant velocity, the temporal profile is the same as the spatial one. Figure 3.17(c) and d show the response of two ON- and OFF-bipolar cells in cats, under the same stimulation. By combining the response profiles of these two cells (ON minus OFF), we obtain rather accurately the same signal as the model's response.

It is worth mentioning that the computations, which have been completed with the MAPLE® symbolic calculation software, needed to be done with 20 significant decimal digits due to the presence of many exponential terms in the mathematical expressions.

This means that, even if we have a good model of the retinal function, a computer simulation cannot be used in real life emulation, not only due to the computational time it would require, but also due to the considerable electrical power needed to run the computer. This fact is in favor of the analogue computation feasible by electronic circuits (see section about the so-called Neuromorphic circuits).

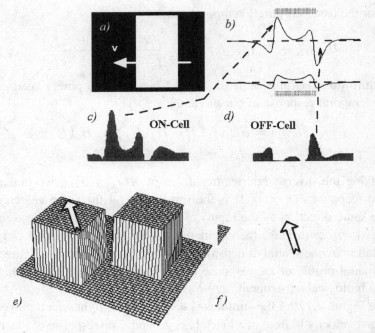

Figure 3.17.   Response to a moving bar. (a) stimulus; (b) spatial profile of the model response; (c) response of an ON cell and (d) response of an OFF cell (after Rödieck and Stone, 1965, with kind permission); (e) 3D view of an intensity image of two squares, one being moving (arrow); (f) model of resulting bipolar cell activity (after Beaudot, 1994).

3.5.3. *Retinal Processing of Moving Images*

In the example of Figure 3.18, the input signal (a) is sequence of a film of a street scene where most of the scene is static; the only motion signal consists of pedestrians (one is crossing the street, two other ones are walking on the pavement).

The output of the bipolar cells model is shown in Figure 3.18(b). Note the spatial high-pass processing: all the surfaces with constant illumination (very low spatial frequencies) appear black whereas the edges of pavement, ground-lines, trees and cars, that are regions of high spatial frequencies, appear brighter. In fact our bipolar cell model gives positive and negative signals, to make them more visible. The image shows the absolute value of the signals.

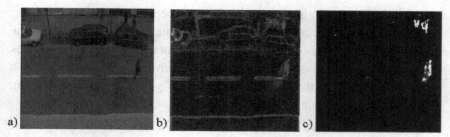

Figure 3.18. Example of the spatiotemporal processing of a scene with moving objects: walking pedestrians. (a) an image of a sequence; (b) the corresponding image at the bipolar cells output; (c) the corresponding output of the Y ganglion cells.

The last part (Figure 3.18(c)) shows the response of the Y or ON-OFF cells. These cells make a double-sided rectification followed by a spatial low-pass filtering and a temporal high-pass filtering. The result is that the static edges of the bipolar layer output are cancelled. Only those corresponding to motion have a temporal component and are likely to be detected. Hence, the functional role of these cells seems to warn the brain about "Where" and "When" something is happening. This Y signal, belonging to the Magno-cellular pathway, could be directly used by the superior colliculus circuits to move the eye in the direction of an event that suddenly happens in the visual field.

3.6. Neuromorphic Circuits

In the domain of engineering applications, the conventional systems of vision are based on the integration of a sensor with one or several digital processors. Because of their high cost and complexity, they mainly concern specific industrial or military applications and do not allow developments toward the field of low-cost and miniature equipment. In fact, they are of little use in on-board applications where small dimensions and low power consumption are needed.

Vision-chips, integrating a powerful (analog or digital) processing at the level of the sensor itself, are good candidates for such applications (Boahen, 1996). These circuits, inspired from the model of biological sensors (Abbott *et al.*, 1996; Jacobs *et al.*, 1996), are studied since the seminal work of Carver Mead (Mead and Mahowald, 1988). They concern over 500 publications in international journals and conferences.

Neuromorphic circuits are part of the concept of "smart sensors", i.e. devices where sensing and processing are intimately linked, and of the same nature (i.e. analog). This concept has many advantages because it suppresses the need for the classical analog-to-digital converter and the need for a temporal sampling of the signals. In the field of vision, these circuits bring many undeniable advantages, but also some draw-backs, as listed below.

3.6.1. *Advantages*

Compared to a system of vision (CCD camera — signal processor), the Vision Chip presents several advantages:

Dynamic range: Most silicon photo-sensors cover a dynamic range of at least 7 decades of luminous intensity. Furthermore, many vision chips are capable of local adaptation, thus widely outperforming conventional circuits that are only able of some global adaptation.

Size: The integration of vision algorithms on silicon insures compact implementation with no possible comparison with conventional systems.

Robustness: The implementation of "collective computation" by analog circuits implies some kind of reliability. In fact, low-pass filters implemented by means of resistive networks (like the basic circuit of the retina) are weakly affected by the mismatch of their components (see Section 3.3.1.2).

Speed: The processing within the various layers is analog and parallel, as well as the transfer between layers. Thus the bottlenecks of sequential transfer and processing between layers in digital systems are avoided. Speed is measured in terms of time constant, and complexity in terms of system order, whatever the number of image pixels. In digital and sequential techniques, the number of pixels proportionally increases the time of computation.

Consumption: Vision integrated circuits are mostly realized in analog C_MOS technology operating in subliminary region. For each elementary operator the energy consumption is low, as well as for the transmission of signals across layers: no bonding pad capacity to charge and very low transient currents.

3.6.2. *Limitations*

Since a series of advantages often involves a series of drawbacks and limitations, we should not ignore the following points:

Reliability of processing: Although the collective character of analog computations, technological constraints linked to the dispersion of the threshold voltages of MOS transistors imply that the desired functions are not always feasible with the required precision. There are solutions but they generally dramatically increase the circuit area dedicated to a given function and lead to a loss of pixel density and to a prohibitive size of the circuit.

Spatial Resolution: The complexity of the feasible treatments is limited by the needed area of circuit. There is always a trade-off between the spatial resolution and the complexity of the processing to be done. One of the largest matrices which have been realized is the one of Andreou and Boahen (1995), its size is 210×230 pixels for 590 000 transistors.

Difficulty of realization: In this domain, there is generally no library of circuits. Each time, for each application, the designer has to create new circuits and, as the technology evolves rapidly toward smaller and smaller dimensions, each time for a same function a new circuit has to be re-designed and the technology has to be characterized for analog functions.

Programmability: There is no general-purpose vision circuit. For each application a new design is necessary. The technique of cellular neural networks (Chua and Yang, 1988) however offers some possibility of programming (Monnin *et al.*, 1998; Monnin *et al.*, 2000).

3.6.3. *Applications*

The first proposed applications were general-purpose artificial retinas able to extract spatiotemporal contrasts (Mead and Mahowald, 1988; Bouvier *et al.*, 1995 and 1996; Funatsu *et al.*, 1995; Mhani *et al.*, 1997) and later to adapt to light intensity (Boahen, 1996; Sicard, 1999), either locally or globally (see Figure 3.19).

After these first steps and the acquired expertise, the designers turned toward higher-level processing tasks. Various kinds of applications have

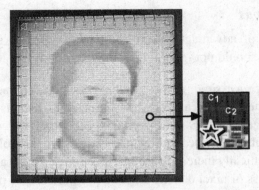

Figure 3.19.   Artificial silicon retina of 64×64 pixels. the insert shows the circuit area dedicated to a pixel: C1 and C2 are the capacitors used in the simulation of receptor and horizontal cells membranes. The star points at the photosensitive area, the rest of the circuit is for resistors and amplifiers. Coll. LIS-CNET Grenoble, France.

been proposed like motion detection (Arreguit *et al.*, 1996), velocity estimation (Delbrück, 1993; Etienne-Cummings, 1997), and even stereovision (Mahowald, 1994).

It is worth noting that these applications are tackled by pioneers who are recognized specialists in the domain of analog microcircuits where research is very active. However, the problems of high-level vision shown in the above examples should be considered as attempts and feasibility studies. In fact, the known digital algorithms cannot be directly translated into silicon circuits. They need a large amount of preliminary theoretical studies, which are missing in the first approaches.

New theoretical studies have been proposed in order to pave the way. For example, in the domain of motion estimation a new architecture has been specified in order to produce a dense estimation of the velocity field (one vector per pixel) with a very good accuracy and a realistic possibility for silicon implementation (Torralba and Hérault, 1999a). In a more general framework the same authors have proposed an analog architecture for two-dimensional filters. These filters can be tuned by linear combinations of asymmetrical oriented filters, leading to non-specific applications of neuromorphic circuits (Torralba and Hérault, 1999b). We also should notice an interesting extension of neuromorphic circuits to the minimization of energy in models of random Markov fields with an application to motion detection (Luthon and Dragomirescu, 1999).

Chapter 4

Neuromorphic Circuits and Motion Estimation

Starting from the mathematical models of moving objects or of ego-motion, in the framework of signal theory and geometrical approach, we will give a review of classical techniques of motion detection and estimation. We will then explore the techniques of the neuromorphic approach, stressing their particular power, mainly due to continuous-time behavior. We will illustrate the dilemma of accuracy-localization of the velocity vector and show the way to handle this problem. We will naturally see in this chapter that some attentional process is mandatory in order to estimate a "good" velocity field.

4.1. Models of Motion

The idea of motion in an image covers several aspects due to different causes. The most general notion of motion is related to the spatio-temporal gradients, *i.e.* the variations of image intensity function $i(x, y, t)$ in time, and in certain directions: temporal and spatial variations are correlated.

This phenomenon may be conditioned by various factors:

- a rigid object (a part of the image), which is translating and/or rotating, see Figure 4.1a.
- a rigid object in the 3-dimensional space which is translating and/or rotating, the retinal projection of which produces added deformations such as skewing or expansion.
- a deformable object which is translating, rotating and expanding (even non-uniformly), see Figure 4.1b. As an example, imagine the motion of smoke, of clouds or of water near the pier of a bridge.

111

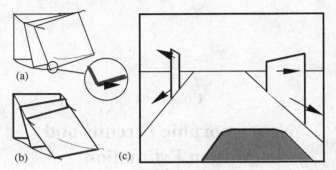

Figure 4.1.   Different kinds of motion. (a) rigid object, (b) deformable object, (c) deformations in the image due to the ego-motion of the observer (for example, the scene of the road while driving a car).

- a shadow or the light reflection of an object passing over other objects of the scene.
- the ego-motion of the observer which makes the projection of rigid objects of the scene change in form with translations, rotations and looming, see Figure 4.1c. Imagine also what a pilot can see when performing trick-flying.
- any combination of two or more of the above-mentioned items, resulting for example from the ego-motion of the observer in a dynamic environment.

Motion estimation is an important domain in computer vision. There are two classes of methods: the energy-based algorithms (in the frequency domain) which provide reliable estimation of velocities and are robust with respect to noise (Watson and Ahumada, 1983; Adelson and Bergen, 1985; Heeger, 1987; Spinei *et al.*, 1998; Bruno and Pellerin, 2002), and the gradient-based algorithms (Horn and Schunk, 1981; Verri *et al.*, 1990; Bouthemy and François, 1993; Barron *et al.*, 1994), which have a lower complexity, at the expanse of sensitivity to noise and aperture problem (see later). Both methods exhibit a high computational demand.

   A powerful solution for real-time processing (as needed in autonomous robots for example) is the realization of specific VLSI circuits, which are more economic in size and power consumption. However, the algorithms for motion estimation on silicon should be adapted from those in computer vision, and they also require a compromise between the number of pixels in the input image and the complexity of each processing unit. In the field of neuromorphic circuits simple motion algorithms have been experimented using analog circuits, see the paper of Sarpeshkar *et al.* (1996) for a review.

Due to the complexity of implementing a bank of spatio-temporal filters, most of VLSI "motion chips" are inspired from gradient-based algorithms (Tanner and Mead, 1988; Deutschmann and Koch, 1998) or correlation-based algorithms (Delbruck, 1993; Kramer, 1996; Deutschmann *et al.*, 1997; Higgins and Koch, 1997) as they allow implementations within very compact micro-circuits. Some other kinds have been proposed with spatio-temporal frequency tuned Gabor filters (Shi, 1998) or space-velocity separable filters (Shi, 1993; Shi *et al.*, 1998).

We will develop in Section 4.2 a neuromorphic architecture implementing a specific frequency-domain method: the velocity-tuned filters inspired from the retinal circuits (Torralba and Hérault, 1999).

4.1.1. *Problems of Motion Estimation*

The estimation of motion in a sequence of images spatially and temporally sampled (for computer vision) or in a continuous time-varying images with spatial sampling (in the case of the retina) consists in the determination of the components of the velocity vector $\{v_x, v_y\}$ at every pixel of the image. The simplest model considers that the movement is due to a translation of the entire image (what happens to your retinal image upon vergence movements). This model may be approximately true for moving objects on a static background if we consider a local analysis of the image, problems will occur at the boarder of the analyzed regions. For larger regions, the model of motion must be more complicated, for example by introducing affine transformations (see later).

In any case, the general problem can be approached classically by two kinds of methods: the frequency domain methods and the spatio-temporal domain methods. For a first step, we will restrict ourselves to the case of continuous signals.

4.1.1.1. *Frequency Domain Approaches*

Let us start with the model of an image submitted to a translation movement. This model has been widely used in the literature and gives interesting results with not too large a computational demand.

Let $i_s(x, y)$ a static image with $I_s(f_x, f_y)$ its 2-D spatial Fourier transform. If we apply to this image in translation motion with horizontal and vertical velocities v_x and v_y, it becomes a spatio-temporal image that can be written as:

$$i(x, y, t) = i_s(x - v_x t, y - v_y t). \tag{4.1}$$

Taking the Fourier transform with respect to spatial variables gives:

$$i_s(x - v_x t, y - v_y t) \xrightarrow{FT_{x,y}} I_s(f_x, f_y)e^{-j2\pi(v_x f_x + v_y f_y)t}.$$

Then, taking the Fourier transform with respect to the temporal variable gives the frequency domain model of a translation motion:

$$I(f_x, f_y, f_t) = I_s(f_x, f_y) \cdot \delta(f_t + v_x f_x + v_y f_y). \tag{4.2}$$

This means that all the energy of the translating image is contained in a plane of the spatio-temporal frequencies domain (Fahle and Poggio, 1981; Adelson and Bergen, 1985), see examples at Figures 4.2 and 4.3, the equation of this plane is:

$$f_t + v_x f_x + v_y f_y = 0. \tag{4.3}$$

In the case of one spatial dimension, the representation is easier and helps understanding. Figure 4.2a shows the perspective view of an image intensity function $i(x - vt)$ with space and time axes, in one spatial dimension. The function is a spatial rectangular pulse translating at a constant velocity, *i.e.* a white band translating in front of a black background.

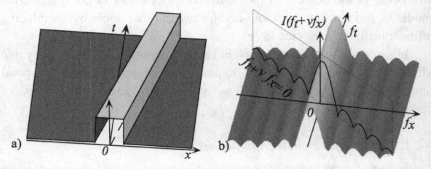

Figure 4.2. Example of a moving 1-D spatial rectangular shape, moving in the spatio-temporal domain (a). Its spatio-temporal frequency spectrum (black curve) as a "cut" of the purely spatial Fourier transform of the rectangle by a "wall of Dirac" (b).

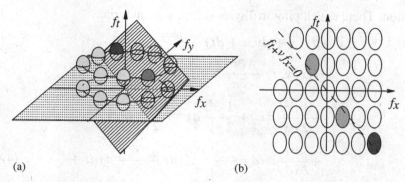

Figure 4.3. (a) the plane of the spatio-temporal frequencies where the spectrum of a translating image is contained. The inclination of the image plane with respect to the horizontal plane gives the amplitude of the velocity vector; the orientation of its intersection with the horizontal plane gives the direction of motion. (b) the much simpler case of one spatial dimension.

A rectangular pulse in space rect(x) has the following form in frequency domain: $\sin(\pi f_x)/(\pi f_x)$. The corresponding frequency spectrum of this moving pulse is given at Figure 4.2b (black curve): it appears as a cut of the purely spatial spectrum $\sin(\pi f_x)/(\pi f_x)$ by a "wall" of Dirac situated in the direction $f_t + v f_x = 0$ of the spatio-temporal frequency co-ordinates.

With two spatial dimensions, the information about direction and intensity of the velocity vector is contained in the orientation of the frequency spectrum, see Figure 4.3 for details.

The method for velocity estimation consists of paving the frequency domain with narrow-band filters tuned to different central frequencies. Then, consider the filters the transfer function of which is intersected by the spatio-temporal frequency plane of the moving image, and compute the energy at the output of each filter: the filters with highest energy will identify the plane (Heeger, 1987), see Figure 4.3 for illustration. Another way is to compute a linear combination of output energies that will give directly the estimated velocity (Spinei *et al.*, 1998).

4.1.1.2. *Spatio-Temporal Domain Approaches*

There is another way of describing translational motion due to Horn and Schunk (1981). The idea is to say that the total derivative of the image intensity function is zero. This means that its only variations are due to motion. Then, developing in Taylor series, we can write:

$$i(x + dx, y + dy, t + dt)$$

$$= i(x, y, t) + \frac{\partial i}{\partial x}dx + \frac{\partial i}{\partial y}dy + \frac{\partial i}{\partial t}dt$$

$$+ \frac{1}{2}\frac{\partial^2 i}{\partial x^2}dx^2 + \frac{1}{2}\frac{\partial^2 i}{\partial y^2}dy^2 + \frac{1}{2}\frac{\partial^2 i}{\partial t^2}dt^2$$

$$+ \frac{\partial^2 i}{\partial x \partial y}dxdy + \frac{\partial^2 i}{\partial x \partial t}dxdt + \frac{\partial^2 i}{\partial y \partial t}dydt + \cdots \quad (4.4)$$

Under the brightness constancy hypothesis (the variations are uniquely due to motion), we have $i(x+dx, y+dy, t+dt) = i(x, y, t)$ obviously, and assuming that the velocity is constant, only the terms with first derivatives have to be considered:

$$\frac{\partial i}{\partial x}dx + \frac{\partial i}{\partial y}dy + \frac{\partial i}{\partial t}dt = 0,$$

and, introducing the velocities dx/dt and dy/dt by dividing by dt:

$$\frac{\partial i}{\partial x}v_x + \frac{\partial i}{\partial y}v_y + \frac{\partial i}{\partial t} = 0, \quad (4.5)$$

or, in vector notation:

$$\nabla i^T v + i'_t = 0,$$

with $\nabla i = (\partial i/\partial x, \ \partial i/\partial y)^T$ the spatial gradient of intensity, $v = (v_x \ v_y)^T$ the velocity vector, and $i'_t = \partial i/\partial t$ the partial derivative of $i(x, y, t)$ with respect to time. Hence, the name "gradient method".

The solution of this equation is obtained directly in the spatio-temporal domain by various procedures after computation of the three partial derivatives of the image. It is interesting to notice that taking the Fourier transform of Equation (4.5) produces an operator that applied to the image Fourier spectrum gives: $j2\pi(f_x \cdot v_x + f_y \cdot v_y + f_t) = 0$, that is the same result: the frequency spectrum of the translating image is all contained in a plan of the spatio-temporal frequencies.

However, there can be noise which corrupts the image or the estimation of the velocities. To make the estimation better, one can try to widen the observation window and take the mean translation vector over it by some smoothing or regularizing technique. But to some extent, this approach can give rise to other problems, because the model of translation motion may be no longer valid in the case of large observation windows.

In this case, the model of motion is no more purely translational: effects of rotations and expansions may occur, and in this case the velocity vectors are no more constant. This means that the higher order terms of Equation 4.4 are no more negligible, which leads to very complex computations.

An interesting trick is to consider an affine model of the velocity together with Equation 4.5, a compromise that, under a less restrictive hypothesis, leads to tractable computations. Here, the velocity vector field $\dot{\mathbf{x}} = (\dot{x}, \dot{y})^T$ is described as a linear transform of the co-ordinates vector $\mathbf{x} = (x, y)^T$ plus a translation velocity $\mathbf{v} = (v_x, v_y)^T$, in matrix-vector representation:

$$\begin{bmatrix} \dot{x} \\ \dot{y} \end{bmatrix} = \begin{bmatrix} a_{11} & a_{12} \\ a_{21} & a_{22} \end{bmatrix} \cdot \begin{bmatrix} x \\ y \end{bmatrix} + \begin{bmatrix} v_x \\ v_y \end{bmatrix}$$

or

$$\dot{\mathbf{x}} = \mathbf{A} \cdot \mathbf{x} + \mathbf{v}. \tag{4.6}$$

Then, returning to the partial derivative equation in spatio-temporal domain,

$$\frac{\partial i}{\partial x} \cdot \frac{dx}{dt} + \frac{\partial i}{\partial y} \cdot \frac{dy}{dt} + \frac{\partial i}{\partial t} = 0,$$

we can write it as: $\nabla i^T \dot{\mathbf{x}} + i'_t = 0$ and insert Equation (4.6):

$$\underbrace{\nabla i^T \mathbf{A} \cdot \mathbf{x}}_{\substack{\uparrow \\ \text{affine} \\ \text{transform}}} + \underbrace{\nabla i^T v + i'_t}_{\substack{\uparrow \\ \text{translation}}} = 0 \tag{4.7}$$

This new partial derivative equation can be solved with the same techniques as for Equation (4.5), see (Cappellini *et al.*, 1993) for a review.

4.1.2. *More General Cases*

4.1.2.1. *The Image May Vary in Time*

Up to now, we have supposed that the motion was that of a *static* image $i_s(x, y)$ submitted to a translation motion. What happens if the image itself is a spatio-temporal function $i(x, y, t)$?

Let be $I_0(f_x, f_y, f_t)$ the 3-D Fourier transform of our original image $i_0(x, y, t)$. By applying a translation movement, the moving image becomes $i(x - v_x t, y - v_y t, t)$. Its spatial Fourier transform with respect to x and y is:

$$I_{x,y}(f_x, f_y, t) \cdot e^{-j2\pi(v_x f_x + v_y f_y)t},$$

and the further Fourier transform with respect to time gives the spectral frequency form of the moving image:

$$I(f_x, f_y, f_t) = I_0(f_x, f_y, f_t) \underset{f_t}{*} \delta(f_t + v_x f_x + v_y f_y), \qquad (4.8)$$

with the convolution product being operated only on the variable f_t. This formula is that of Equation (4.2), with the simple product being replaced by a convolution product. By realizing the operation of convolution, we obtain the final representation of the moving image:

$$I(f_x, f_y, f_t) = I_0(f_x, f_y, f_t + v_x f_x + v_y f_y). \qquad (4.9)$$

The spectrum of the moving image lies no longer on a plane of equation $f_t + v_x f_x + v_y f_y = 0$ in the spatio-temporal frequency the co-ordinates. Figure 4.4 gives an example of this transformation between the spectra of original and moving images, in the simpler case of only one spatial dimension. Four important and curious facts have to be noted:

- Figure 4.4a: because the original image is time-dependent, its spectrum has obviously some temporal frequency extent. A static image (constant

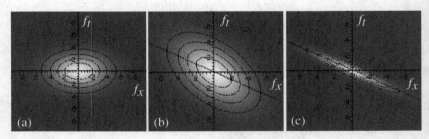

Figure 4.4. Spatio-temporal spectra (one spatial dimension): (a) of the original time-varying image, (b) of the image moved at a velocity V_1, (c) same thing, but if the image would have been of narrower temporal spectrum. Dotted line: the line of equation $f_t + v f_x = 0$.

in time) would have a spatio-temporal frequency spectrum lying on a Dirac wall at temporal frequency 0.

- Figure 4.4b: the spectrum of the moving image seems (*only seems*) to rotate in a direction $f_t + vf_x = 0$ compatible with motion, BUT its axis is NOT collinear with the motion direction as would be expected. In fact, we should not be surprised: our new hypothesis is not that of Section 4.1.1.1, hence temporal variations are not only due to motion.
- As a consequence, integrating the power spectrum (*i.e.* computing the energy) on a test line of equation $f_t + v_{test} f_x = 0$ would not give a maximum when $v_{test} = v$.
- Figure 4.4c: a quite surprising case! If the original image would have been of narrower temporal spectrum extent, its moving spectrum would have been more coaxial with the motion direction.

If the second observation is true, the third one is FALSE! Let us verify by computing the energy E_v of the moving image along a line of equation $f_t + v_{test} f_x = 0$. For this, we integrate the spectral distribution of energy (here: the square of the amplitude spectrum) along the line:

$$E_{v_{test}} = \iint_{f_x, f_t} |I_0(f_x, f_t + vf_x)|^2 \cdot \delta(f_t + v_{test} f_x) \, df_x df_t,$$

that is:

$$E_{v_{test}} = \int_{f_x} |I_0(f_x, (v - v_{test}) f_x)|^2 \, df_x. \tag{4.10}$$

First, let us remember that the frequency spectrum of images decreases if either spatial or temporal frequency increases (remember the $1/f$ shape). Second, let us note that the term $(v - v_{test}) f_x$ replaces f_t in the original expression of $I_0(f_x, f_t)$; that is when the coefficient $(v - v_{test})$ is larger, the profile of the original $I_0(f_x, f_t)$ is narrower in the direction of f_t, reducing the area of the profile. Hence, the integral of Equation (4.10) will be maximum when this coefficient is 0, that is when $v_{test} = v$. Figure 4.5 illustrates the variation of E_v when v_{test} varies.

This is a counter-intuitive result regarding Figure 4.4; it demonstrates that the hypothesis showing that the image translated under motion should be of a static nature in order to extract velocity, is not mandatory.

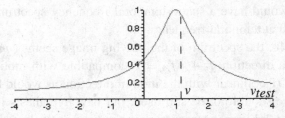

Figure 4.5. Relative variation of the energy according to the value of test velocity. The maximum is obtained for $v_{test} = v$.

If the test function, instead of being a Dirac (infinite length and zero width) oriented at the velocity v_{test}, is a low-pass spatio-temporal filter, that is, with a finite length and a non-zero width, the same result is obtained. We will call such a filter a "velocity-tuned filter" (VTF).

4.1.2.2. *Motion Transparency*

A majority of researches on visual motion processing have been concerned with the measure of a unique velocity vector in a local region of the image. However, psychophysical experiments have shown that multiple motions can be perceived in a single region of the visual field. Multiple motions happen in various cases, for example a moving shadow cast on a stationary surface, a moving object partially occluded by foliage in the wind, or a moving object seen by reflection when looking through a glass window (Smith *et al.*, 1999). These cases of motions are called *motion transparencies*. The process can be additive (glass window) or multiplicative (cast shadow). Neurophysiologists have shown that motion transparency is processed in area MT rather than in area V1: the visual system computes the global motion of objects and regions, mainly in extrastriate brain areas as V5/MT from the signals of striate cortex V1.

In the case of motion transparency, the existence of two or more velocity vectors at the same image location, require some adaptation of classical techniques (Pingault *et al.*, 2003). In the spatio-temporal frequency representation, multiple motions generate several planes of motion instead of one (imagine the representation of Figure 4.3 with several planes of motion!). Then, it is mandatory to pool the activity of many elementary motion detectors in order to solve the problem. An accurate solution has been recently proposed (Pingault and Pellerin, 2004), by estimating the maxima of the local Fourier spectrum of the image sequence.

4.1.2.3. *Influences of Spatial and Temporal Sampling*

In the retina, the 2D space is sampled by the photoreceptors, but the processing in time is continuous. In CCD cameras, both space and time are sampled, the consequences are quite different in both cases. Let us first consider the retina.

To make it simpler, we will consider spatio-temporal images in one spatial dimension. It is known in signal processing that when a continuous signal is sampled, its Fourier transform becomes periodic. Thus, taking the example of Figure 4.2, the input image (projected onto the retina) is a continuous spatial rectangular pulse of width w, moving at a constant velocity v: $i_c(x, t) = \text{rect}_w(x - vt)$. Its Fourier transform is:

$$I_c(f_x, f_t) = \frac{\sin(\pi f_x w)}{\pi f_x} \delta(f_t + v f_x) = w\,\text{sinc}(f_x w),$$

and its representation is given at Figure 4.2b.

Under spatial sampling every Δl, the sampled image becomes:

$$i_s(x, t) = \text{rect}_w(x - vt) \cdot \sum_{k=-\infty}^{+\infty} \delta(x - k\Delta l), \qquad (4.11)$$

$\sum_{k=-\infty}^{+\infty} \delta(x - k\Delta l)$ is called a "Dirac comb". As shown at Figure 4.6, the periodization of a function with a non-limited support produces "aliasing", in other words, the original continuous function is not recoverable from its sampled version.

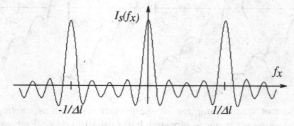

Figure 4.6.   Fourier spectrum of a sampled rectangular pulse. Observe the periodicity.

The Fourier transform of a simple product in space is the convolution product in frequency and the Fourier transform of a Dirac comb of pitch Δl is a Dirac comb in frequency of pitch $1/\Delta l$. Hence, the spatio-temporal Fourier transform of our sampled image is:

$$i_s(k\,\Delta l,\,t) \xrightarrow{F_{x,t}} = [w \operatorname{sinc}(f_x w) \cdot \delta(f_t + v\,f_x)] * \frac{1}{\Delta l} \sum_{n=-\infty}^{+\infty} \delta\left(f_x - \frac{n}{\Delta l}\right),$$

or

$$I_s(f_x,\,f_t)$$

$$= \frac{1}{\Delta l} \sum_{n=-\infty}^{+\infty} \left[w \operatorname{sinc}\left(\left(f_x - \frac{n}{\Delta l}\right)w\right) \cdot \delta\left(f_t + v\left(f_x - \frac{n}{\Delta l}\right)\right)\right],$$

$$(4.12)$$

that is, the drawing of Figure 4.2b repeated every $1/\Delta l$ along the spatial frequency axis as shown in Figure 4.7.

It is important to observe here that, contrarily to Figure 4.6, the periodization does not produce aliasing, every repeated pattern is isolated and fully recognizable: it is possible to reconstruct from it the original continuous image without any loss of information.

In CCD cameras, we also have a spatial sampling, and the same property of periodization in spatial frequency. But, in addition, we have a time

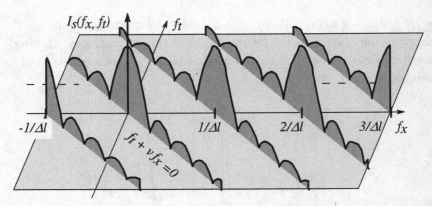

Figure 4.7. Effect of spatial sampling on the Fourier spectrum of moving image. The spectrum of the continuous image is repeated, but no aliasing occur, each pattern is easily identifiable.

sampling (50 or 60 frames per second, according to the frequency of the power supply. Here is the beginning of our trouble: just look at Figure 4.7 and imagine a repetition of its entire pattern along the temporal frequency axis with some given periodicity. For some values of the velocity, the patterns of the original continuous spectrum will overlap and produce aliasing.

This is another advantage of the biological systems over Man-made technical systems: the continuous time associated to motion avoids the problems of aliasing. It may be the reasons why motion is so important to vision.

It is interesting to recall how, in old TV systems, under-sampling was used in order make a person not recognizable. In fact, when the subject was still, we saw only gray squares on the screen, but when the subject was moving his (or her) head, the features of the face became recognizable. Nowadays, the problem is solved by simply adding noise to the under-sampling.

4.2. Velocity-Tuned Filter and Motion Estimation

4.2.1. *Principle*

If we recall the spatio-temporal transfer function of the basic retinal circuit (Chapter 3, Sections 3.2.1 and 3.2.2), we notice that it is a filter tuned to zero-velocity images: the axes of the transfer function lie on the frequency axes f_x and f_t.

From what has been seen in the preceding section, we can easily imagine how to transform it into a filter tuned to velocity v_0 by replacing f_t by the typical term $f_t + v f_x$ in the transfer function, in one spatial dimension:

$$\frac{1}{1 + 2\alpha\,(1 - \cos(2\pi f_x)) + j2\pi\,\tau f_t}$$

$$\Rightarrow \frac{1}{1 + 2\alpha\,(1 - \cos(2\pi f_x)) + j2\pi\,\tau\,(f_t + v_0 f_x)},$$

or in two spatial dimensions:

$$H_{\mathbf{v}}(f_x, f_y, f_t) = \frac{1}{\begin{aligned}&1 + 2\alpha\,(2 - \cos(2\pi f_x) - \cos(2\pi f_y))\\&+ j2\pi\,\tau\,(f_t + v_{0x} f_x + v_{0y} f_y)\end{aligned}}, \qquad (4.13)$$

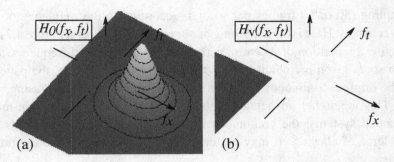

Figure 4.8.   Velocity-tuned filter (a) the basic retinal circuit transfer function: the axes lie on f_x and f_t. (b) the transfer function of the velocity-tuned filter.

which represents a velocity-tuned filter (VTF), tuned to the velocity $\mathbf{v}_0 = (v_{0x}, v_{0y})^\mathsf{T}$. Figure 4.8 gives an example in one spatial dimension.

How can we design an electrical circuit able to present this transfer function?

4.2.1.1. *Equation and Structure*

Let us take only one spatial dimension in Equation (4.13) for simplicity:

$$H_v(f_x, f_t) = \frac{1}{1 + 2\alpha\left(1 - \cos(2\pi f_x)\right) + j2\pi\tau\left(f_t + v_0 f_x\right)}, \quad (4.14)$$

and consider that the term $j2\pi\tau(f_t + vf_x)$ at denominator is, in the Fourier domain, the equivalent of the continuous derivative with respect to t and to x. It is fully justifiable for the continuous time variable, but not for the sampled spatial variable. We should replace the part $j2\pi\tau vf_x$ by the equivalent of the symmetric derivative in the sampled domain:

$$\frac{\partial i}{\partial x} \approx \frac{i((k+1)\Delta l, t) - i((k-1)\Delta l, t)}{2\Delta l}$$

or, with the convention $\Delta l = 1 : \partial i/\partial x \approx (i(k+1, t) - i(k-1, t))/2$.
The equivalent of this approximation in the Fourier domain is:

$$\frac{\partial}{\partial x} \xrightarrow{F} j2\pi f_x \approx \frac{e^{j2\pi f_x} - e^{-j2\pi f_x}}{2} = j\sin(2\pi f_x).$$

Hence, Equation (4.14) becomes:

$$H_v(f_x, f_t)$$

$$= \frac{1}{1 + 2\alpha\,(1 - \cos(2\pi f_x)) + j2\pi\,\tau f_t + jv_0\tau\,\sin(2\pi f_x)}. \quad (4.15)$$

Now, in order to derive the differential equation which describes the circuit, let us develop the sine and cosine functions into exponential functions:

$$H_v(f_x, f_t)$$

$$= \frac{1}{1 + 2\alpha - \alpha\left(e^{j2\pi f_x} + e^{-j2\pi f_x}\right) + j2\pi\,\tau f_t + \frac{v_0\tau}{2}\left(e^{j2\pi f_x} - e^{-j2\pi f_x}\right)},$$

then, by factorizing the exponential terms of same sign:

$$H_v(f_x, f_t) = \frac{1}{1 + 2\alpha - \alpha(1 - \frac{v_0\tau}{2\alpha})e^{j2\pi f_x} - \alpha(1 + \frac{v_0\tau}{2\alpha})e^{-j2\pi f_x} + j2\pi\,\tau f_t}.$$

By replacing $H_v(f_x, f_t)$ by the output/input ratio $O(f_x, f_t)/I(f_x, f_t)$ and taking the inverse Fourier transform, we obtain the difference-differential equation of the filter:

$$i(k, t) - o(k, t) = 2\alpha o(k, t) - \alpha\left(1 - \frac{v_0\tau}{2\alpha}\right)o(k + 1, t)$$

$$- \alpha\left(1 + \frac{v_0\tau}{2\alpha}\right)o(k - 1, t) + \tau\frac{do(k, t)}{dt},$$

and, by replacing α and τ respectively by r/R and r/C:

$$\frac{i(k, t) - o(k, t)}{r} = \frac{o(k, t) - \left(1 - \frac{v_0\tau}{2\alpha}\right)o(k + 1, t)}{R}$$

$$+ \frac{o(k, t) - \left(1 - \frac{v_0\tau}{2\alpha}\right)o(k - 1, t)}{R} + C\frac{do(k, t)}{dt},$$

that is the equation of a Resistor-Capacitor line with isolating amplifiers in the horizontal connections, as shown in Figure 4.9. The gain A of the amplifiers is different according to their orientation.

The amplifiers oriented leftward have a gain $A = 1 - v_0\tau/2\alpha$, and those oriented rightward have a gain: $A = 1 + v_0\tau/2\alpha$. According to the sign of the tuning velocity v, the gain is higher or lower than 1. We should

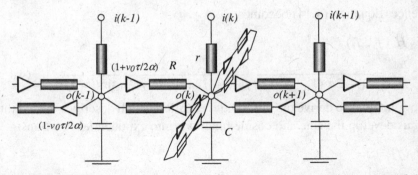

Figure 4.9.   Circuit for a velocity-tuned filter. Observe the asymmetric structure of the coupling between cells.

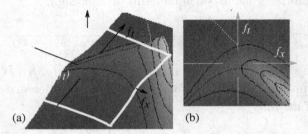

Figure 4.10.   The VTF's spatio-temporal transfer function: (a) Perspective view, (b) Top view. Note the influence of the sine term with respect to Figure 4.8b.

consider that the sum of the two gains is constant and equal to 2. This circuit can be generalized to the case of two spatial dimensions by simply adding transverse couplings as shown at the center point in Figure 4.9.

4.2.1.2. *The Spatio-Temporal Response*

Figure 4.10 represents the spatio-temporal transfer function of the velocity-tuned filter. The crest line of the graphic (top view) shows clearly the influence of the term in sine at the denominator of Equation (4.14). It results from the discrete approximation of the continuous derivative.

To derive the impulse response, we simply take the inverse Fourier transform of Equation (4.15). The result is given by:

$$h_v(x, t) = \frac{1}{2\sqrt{\pi \tau \alpha t}} e^{-\frac{t}{\tau}} e^{-\frac{(x - v_0 t)^2 \tau}{4t\alpha}},$$

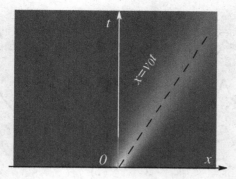

Figure 4.11. The VTF's spatio-temporal impulse response (top view).

or, in two spatial dimensions:

$$h_v(x, y, t) = \frac{1}{4\pi \alpha t} \, e^{-\frac{t}{\tau}} \, e^{-\frac{(x-v_{0x}t)^2 \tau}{4t\alpha}} \, e^{-\frac{(y-v_{0y}t)^2 \tau}{4t\alpha}}. \qquad (4.16)$$

Figure 4.11 shows the impulse response, in one spatial dimension, of a filter tuned to velocity v_0: the response propagates in the direction of the velocity while being gradually attenuated.

4.2.1.3. *Response to Moving Patterns*

In order to have an idea of which parameter of the response will be important, let us look at the spatial profile of the filter's response to a moving pulse (Figure 4.12a. The filter is tuned to a velocity v_0 and is stimulated by a moving impulse at velocity v. When $v = v_0$, the amplitude of the response is maximal and its profile is symmetric. When velocity v is in the same direction as v_0 and slightly higher in value, or in the opposite direction (with any value), the response's amplitude is lower and its profile shows a trailing edge opposite to the direction of motion, exhibiting causality or anti-causality. Moreover, if we compute the energy of the response by spatially integrating its squared value, it is clear that the output energy is a simple and interesting measurement of the response.

The influence of the circuit's parameters can also be shown. The (one-dimensional) spatial profile of the response is given by the following

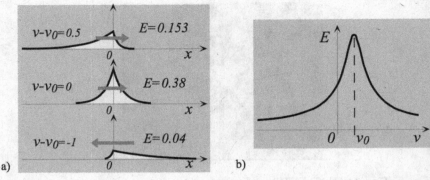

Figure 4.12. VTF's response: (a) spatial profile of the response to a moving impulse. Top: rightward, with a higher velocity than optimal one. Middle: rightward with optimal velocity. Bottom: leftward. (b) The output energy E is maximal when v equals the optimal velocity.

formula:

$$r(x) = \frac{e^{-\frac{\sqrt{4\alpha + \tau^2(v - v_0)^2}|x|}{2\alpha}}}{\sqrt{4\alpha + \tau^2(v - v_0)^2}} e^{-\frac{\tau(v - v_0)x}{2\alpha}} \qquad (4.17)$$

The first term is symmetric in x, the second one is responsible for the asymmetry according to the difference $(v - v_0)$. The time constant τ acts as a proportionality factor to this difference, and the space constant α has some influence on the width of the profile.

4.2.2. Circuit for Motion Estimation

4.2.2.1. The Necessity of a Pre-Filtering

We have seen that the fundamental principle is to estimate the plane of maximal energy in the spatio-temporal frequency domain. But this operation is not always easy. The shape of the frequency spectrum depends on image content: in the textured regions with many details, the spectrum is sufficiently wide to allow the recognition of a plane, but in homogeneous regions, the spectrum is very narrow and localized in the vicinity of low spatio-temporal frequencies. To overcome the problem of this last case, it is necessary to "widen" the spectrum. This can be done by applying a high-pass filter that compensates for the $1/f$ shape of the spectrum: in signal processing, this operation is called "spectral whitening". This way, the frequency spectrum is wide enough to identify the existence of a plane. Amazingly, this high-pass filtering is precisely what is done by the outer plexiform layer of the retina (Chapter 3, Section 3.3.1.2), just as if Nature had identified this problem.

Thus, before any operation of motion estimation, we will apply a retinal filter in order to whiten the spatio-temporal signal.

4.2.2.2. *Estimation of Local Energy*

According to Figure 4.12, two properties of the VTF's response may be used to estimate velocity: the spatial shape of the response and the output energy of the filter. The spatial shape is not relevant because it varies according to the image's shape. The output energy is much more interesting because it is largely independent on the image's shape.

This energy is estimated by integrating the filter's output in the spatio-temporal signal domain or, equivalently in the spatio-temporal frequency domain. In this frequency domain, the filter's output is the product of the transfer function of the filter by the Dirac plane of the motion. And the output energy will be obtained by integrating the output's energy spectral density function over the three frequencies axes. For energy-bounded signals like images, the energy spectral density function is merely the squared amplitude of the complex Fourier transform of the signal. Hence, the output energy takes the form:

$$E := \iiint_{f_x, f_y, f_t} \left| H_{\mathbf{v}_0}(f_x, f_y, f_t + \mathbf{v}_0^\mathsf{T}\mathbf{f}) \right|^2 \left| I(f_x, f_y, f_t) \right|^2$$
$$\times \delta(f_t + \mathbf{v}^\mathsf{T}\mathbf{f}) df_x df_y df_t,$$

where $\mathbf{v}^\mathsf{T}\mathbf{f} = v_x f_x + v_y f_y$ for the image velocity and $\mathbf{v}_0^\mathsf{T}\mathbf{f} = v_{0x} f_x + v_{0y} f_y$ for the optimal velocity of the VTF.

The first term in the integrand is the filter's transfer function amplitude, the second one is the spectral density function of the image's energy and the third one is the Dirac plane of motion. Because of the integration with a Dirac function this three-dimensional integral reduces to a two-dimensional one:

$$E = \iint_{f_x, f_y} \left| H_{\mathbf{v}_0}(f_x, f_y, (\mathbf{v}_0 - \mathbf{v})^\mathsf{T}\mathbf{f}) \right|^2 \left| I(f_x, f_y, -\mathbf{v}^\mathsf{T}\mathbf{f}) \right|^2 df_x df_y.$$

$$(4.18)$$

Figure 4.12b has been drawn from this expression. In fact, this formula gives the *total* energy of the filter's output. It is equivalent to the estimation of the total energy by integrating in the spatio-temporal domain over the *whole* image. But in our purpose, we rather need a *local* estimation of the energy because we want to estimate a local velocity vector. Hence, the integration should be processed only on a restricted spatial domain, where the velocity vector has to be estimated. In practice, the estimation of the local energy $E_L(x, y, t)$ at position (x, y) and at time t will be done by i) squaring the filter's output $r(x, y, t)$ (like in Equation 4.17), and ii) integrating it over a weighting window, the estimation window $w(x, y)$:

$$E_L(x, y; t) = \iint_{x,y} |r(x, y, t)|^2 \, w(x, y) dx dy.$$

The width of the estimation window is of first importance. It will be chosen according to specific conditions. In principle, the estimated velocity will be more accurate with a narrower window, however in case of noisy signals, it would be interesting to widen the estimation window in order to have a mean effect (Braddick, 1993). We will see later that this widening is also mandatory in the case of the so-called "aperture problem" and when the camera is moving (case of ego-motion).

4.2.2.3. *Combination of VTF*

Now, for each velocity-tuned filter, we know the local energy. How is it possible to find the velocity vector. A solution could be to take a large number of selective VTFs, each of them tuned to a particular velocity. The optimal velocity of the filter with the highest energy would be the searched velocity. This solution could give accurate results, only if we had a large number of filters, which would be very expensive in term of circuit or in term of calculation.

We will prefer a solution widely used in nature: a limited number of loosely tuned filters. This solution is also used in radio-engineering to discriminate frequency modulation: take the difference of the signals from two filters tuned to two neighboring frequencies. We will get inspired from this solution. We take two filters respectively tuned to velocities $+v_0$ and $-v_0$, and compute the difference between their output energies E_+ and E_-. Because we want to obtain a result independent of the input signal, we normalize this difference by the sum of the output energies. Hence, the

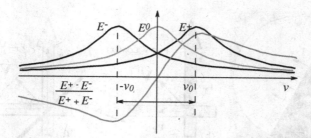

Figure 4.13. Combination of two VTFs. The difference of output energies normalized by their sum provides a linear estimation of the velocity between $-v_0$ and $+v_0$.

estimated velocity is given by:

$$\hat{v} = \frac{E_+ - E_-}{E_+ + E_-}. \tag{4.19}$$

Figure 4.13 gives the graph of estimated velocity from Equation (4.19). We observe a quasi-linear variation between the two limits of filters' optimal velocities $-v_0$ and $+v_0$. This simple combination will be usefull if we choose v_0 as the absolute value of the highest possible velocity in the image. A better formula (Equation 4.20) provides full linearity by introducing a third filter tuned to optimal velocity zero:

$$\hat{v} = \frac{1/E_-^2 - 1/E_+^2}{1/E_+^2 + 1/E_-^2 - 2/E_0^2}. \tag{4.20}$$

Of course it is obtained at the expanse of a more complex circuitry or longer calculation (Torralba and Herault, 1999). The idea of combining three VTF has been also successfully used with spatio-temporal Gabor filters by Spinei and Pellerin (2001) for motion estimation.

4.3. The Aperture Problem

4.3.1. *Definition*

In many situations, it is not adequate to consider motion at a purely local level. Suppose that we are looking at the image through a small circular aperture, and that the observed region is a black square larger than the aperture, moving in some direction (white arrow at Figure 4.14a).

The only perceived motion is in a direction orthogonal to the side of the square, whatever the real motion is (Adelson and Movshon, 1982).

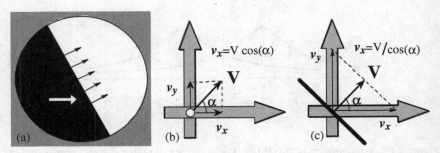

Figure 4.14. The aperture problem: (a) a large edge is seen trough a small window, if it moves, the perceived motion is orthogonal to the edge, whatever the direction of the real motion (white arrow). (b) if a dot is moving, the measured components of its velocity are the orthogonal projections of the motion vector on the directions of motion sensors. (c) if a line is moving, the measured components are parallel projections of its orthogonal apparent motion.

This effect is called "aperture problem", it occurs when the integration is processed over a too narrow estimation window. The directionally selective neurons of area V1 (with their small, oriented receptive fields) are likely to provide such ambiguous motion measurements. To derive the true velocity, the integration should be performed over a wider field. Such a process is believed to happen in neurons of area V5 or MT in the primate's brain (Movshon *et al.*, 1986).

In two spatial dimensions, let us suppose that we have two sets of VTF, one to estimate the horizontal component of the image velocity, and one to estimate the vertical component (Figure 4.14b and c). It can be shown that:

- for a dot moving with a velocity vector **v** of amplitude V and direction θ, the two measured components are respectively $v_x = V\cos(\theta)$ and $v_y = V\sin(\theta)$,
- for a moving line with an orthogonal velocity vector **v** of amplitude V and direction θ, the two measured components are respectively $v_x = V/\cos(\theta)$ and $v_y = V/\sin(\theta)$.

4.3.2. Detection of Aperture Problems

It is normally admitted that two sets of orthogonal velocity filters are sufficient to estimate motion. Following Torralba and Herault (1999), let us consider what happens with four sets of VTF.

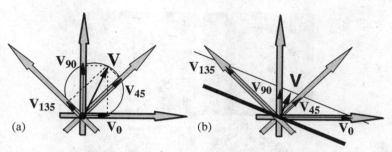

Figure 4.15. Detection of the aperture problem with a set of 4 VTF (a) in normal conditions, all the estimated components lie on a circle. (b) in the case of aperture problem, the components lie on a line.

In the case of non-ambiguous motion (i.e. dot in Figure 4.15a), all the projections of the real velocity vector on the elementary VTFs are situated on a circle, the diameter of which corresponds to the velocity vector to be estimated. In the case of ambiguous motion (aperture problem), all the estimated velocities lie on a line parallel to the moving contour as in Figure 4.15b.

4.3.3. *First Solution: Spatial Integration*

This simple trick allows us to determine whether we are dealing with an aperture problem or not. The solution seems now easy to derive: if there is no aperture problem, the integration window should be small to get the best accuracy, and if the aperture problem is present, the integration window should be enlarged until the problem disappears.

Figure 4.16 illustrates this fact: a ring is moving from top left to bottom right. When the integrating window is narrow, only the orthogonal velocity is estimated and the ring should be perceived as elastic and changing in form. When the integrating window is sufficiently large, the problem disappears and the correct velocity vectors are estimated. They are all parallel and the ring is perceived as rigid.

However, this is far from being sufficient. Because the integration window is large, new problems may happen in the presence of multiple objects moving in different directions.

Figure 4.16c and d illustrates this new problem: another ring is present and moves in the opposite direction of the first one. With a small integration

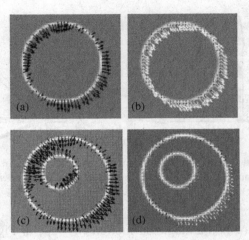

Figure 4.16. First solution to the aperture problem: a ring is moving, (a) with a small integration window, only the orthogonal motion is estimated. (b) with a large integration window, the right motion vector is estimated. If two rings are moving in opposition, in (c) with a small integration window, the same problem as in (a) appears. In (d) with a large integration window, the motion is cancelled in the vicinity of the two rings.

window, the same problem as in Figure 4.16a is present, and when the integration window is large, local motion vectors are summed and neighboring portions of the rings seem to be still. The full solution to motion estimation requires more in-depth considerations.

4.3.4. *Solution with Limited Spatial Integration*

In fact, the same problem will be present at the boundaries of objects. If the object moves while the background remains still, even if there is no aperture problem, the motion at the boundary will be under-estimated because of fusion with the zero-velocity background.

The solution is simple: instead of integrating symmetrically in all directions around a point (circular estimation window), we should perform the integration over continuous regions and stop it at the borders or contours, where we are supposed to have a transition between an object and another one, or between an object and the background.

Figure 4.17 illustrates this solution. In Figure 4.17a, the integration is preformed by a resistive network, the space constant of which gives the extent of the estimation window. If we "cut" the resistors of this network

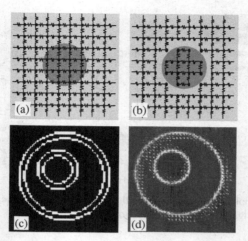

Figure 4.17. Second solution to the aperture problem. (a) and (b) the integration process should be "cut" at the object's boarder and applied separately in each defined regions. (c) borders of the two rings. (d) with a separation of the integration regions, the estimated motion is the right one.

at the contours of the object (Figure 4.17b), there will be two operations of integration; one within the object, and one outside, each estimating the mean velocity of its associated region. As an example, Figure 4.17c shows the contours of the two rings of Figure 4.16, delimiting 5 regions of integration, each having its own motion. Figure 4.17d shows the resulting estimated velocity vectors: both rings' motions are correctly estimated and look both rigid.

Here we have a good example of the interest of contours in an image. They help discrimination between the attributes of neighboring regions. Here, again, we are dealing with a property of the retina: the spatial high-pass filter of parvocellular ganglion cells helps to enhance contours.

4.4. Application Examples

4.4.1. *Circuit Example for One Spatial Dimension*

4.4.1.1. *Architecture of the Circuit*

In order to better understand the process of motion estimation, we first illustrate the case of one spatial dimension. Figure 4.18 gives the global

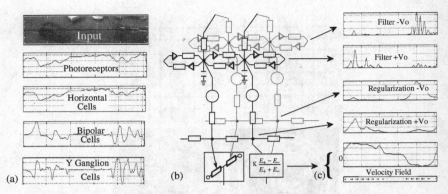

Figure 4.18. Example for 1-D motion (a) the retinal signals, (b) the circuit for motion estimation, (c) the signals and estimated motion.

architecture and the associated signals. The input is one row extracted from an image sequence of a cross-road where vehicles are moving (black line at top of Figure 4.18a).

The first step of processing is realized by a retinal circuit in order to whiten the input signal. We see at Figure 4.18a the various signals of the retinal model: photoreceptors, horizontal and bipolar cells' outputs and the output of Y ganglion cells. The outputs of bipolar cells exhibit the whitened version of the input signal, and the Y cells layer brings a further temporal high-pass filter, which is relevant for motion estimation because it eliminates the static regions.

Now, the input signal has been sufficiently pre-processed for motion estimation. Figure 4.18b gives the architecture of the system: One circuit (in black) for estimating motion towards right $(+v_0)$ and one circuit (in gray) for estimating motion towards left $(-v_0)$, and a final stage combining the outputs of both circuits to estimate the velocity.

At top, we recognize the circuits of the two velocity-tuned filters with their asymmetrical structure. Their outputs are squared (circles) in order to provide the local energy and spatially low-pass filtered by a smoothing resistor network in order to reduce noise and to solve the aperture problem. The last box at bottom computes the velocity. It is realized by a non-linear circuit: two resistors R_1 and R_2 in series between two voltage generators V_1 and V_2. The resistors' values are modulated by the smoothing network outputs and the result u is the voltage at the common point:

$$u = \frac{V_1 R_2 + V_2 R_1}{R_1 + R_2}.$$

If $V_2 = -V_1$, with $R_1 = \frac{k}{E_+}$ and $R_2 = \frac{k}{E_-}$, we obtain: $u = V_1 \frac{E_+ - E_-}{E_+ + E_-}$, that is the desired velocity estimation.

4.4.1.2. *Signals and Estimated Motion*

Figure 4.18c shows the signals: the top two diagrams show the local energy (squared outputs of VTF's), the following two diagrams are the smoothed (regularized) versions of the local energies and the last one is the estimated velocity. We see that the car on the left-hand side runs rightwards and the truck behind a tree on the right-hand side runs leftwards.

We also notice on this last diagram that the spatial variation of the estimated velocity does not present abrupt fronts at the contours of moving objects as it would be expected. This is due to the smoothing process which here does not break at contours, giving a graded response between the moving objects and the still background.

4.4.2. *Examples with Two Spatial Dimensions*

4.4.2.1. *The Camera is Static*

The problem is typical: detecting moving objects on a static background, the camera being also static. The example of Figure 4.19 deserves attention.

In this figure, a car is approaching towards the camera, the sunlight creates a shadow which is attached to the moving object. In (a) we have an image of the sequence, in (b) we see the signal of the bipolar cells and in (c) its temporal high-pass filtered version. In (d) the optical flow is shown only every 5 pixel for clarity (in fact, it is available at each pixel). We notice that:

- the global motion is not the translation of a rigid object but a looming effect: the velocity field exhibits an expansion,
- the shadow on the ground seems to belong to the moving object.

Figure 4.19. Example for 2-D motion. (a) input image, (b) image of bipolar cells, (c) image of Y ganglion cells, (d) optical flow. *Note*: the shadow also is moving.

In this last case, we are dealing with a problem of multiplicative motion transparency. Our process is not able to tell that there are two velocity vectors at each pixel of the shadow, one moving at the object's speed and one with zero velocity (the ground).

4.4.2.2. *The Camera is Moving*

Another kind of problem appears when the camera itself is moving. Then, the optical flow theoretically occupies the whole image: to get an idea of the motion field, imagine what you see when it is snowing and you are driving in the night: the trajectories of snowflakes meet exactly the optical flow.

This kind of motion is called "egomotion" (Dahmen *et al.*, 1997; Matthias *et al.*, 1998). In this case, if the whole visual scene is composed of texture, every pixel is moving and the velocity field is dense. But for example in the case of Figure 4.20, a tree in front of a uniform background, a motion vector is only present on the textured object (*i.e.* the branches of the tree). However, it is evident that in the case of egomotion, regardless of

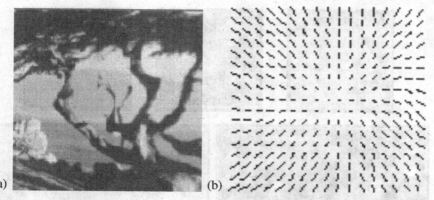

Figure 4.20. Example of ego-motion: (a) the input image, and (b) the optical flow. *Note*: the velocity field concerns the whole image, velocity vectors are estimated even in open regions with no object.

the scene, the velocity field should be of the same type and should virtually occupy the whole image.

Hence, knowing that we are moving, we can widen the integration window of our VTF, so that a motion vector can be present at every pixel. This is what has been done in Figure 4.20c. Under this condition, if an object is simultaneously moving in the scene, it can be identified by subtracting the local motion from the global estimated flow field.

It is also important to notice that the analysis of the flow field during egomotion is able to provide information about the 3D structure of the scene itself (Ramachandran *et al.*, 1988; Cirrincione, 1998; Cirrincione, and Cirrincione, 1999).

4.4.2.3. *Using Objects Boundaries*

As stated in Section 4.3.4, the process of integration should be spatially limited in order to estimate the correct velocity of moving objects. Figure 4.21 illustrates the problem in a real case. The image sequence (known as "taxis of Hamburg") is very complex because the camera is not fully static: it is subject to vibrations, which give an overall small motion, whereas the objective is to estimate the motion of the street traffic.

First, the process must comprise some thresholding operation to remove the small motion components, and then estimate the velocity of moving

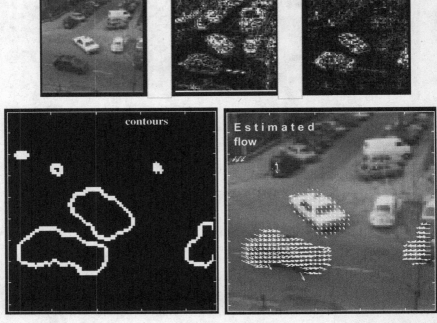

Figure 4.21.   Example for 2-D motion with object boundaries. *Top*: input image, image of bipolar cells and image of Y ganglion cells. *Bottom left*: optical flow. *Bottom right*: motion vectors on moving objects.

objects. Second, because a simple integration through a circular window would smear the velocity field around objects, a detection of contours will be used to "cut" the spatial integration process on contours, as described in Section 4.3.4. This results in a correct estimation of motion inside each object.

Chapter 5

Color Processing in the Retina

Most of the functional models of the retina present the effect of chromatic opponency as an intrinsic property of the retinal circuitry (Guth and Lodge, 1973; Paulus and Kröger-Paulus, 1983; De Valois and De Valois, 1993; Dacey, 1996; Calkins and Sterling, 2002; Teufel and Wehrhahn, 2004). This idea gives comfort to many authors as it deals with the observation of this effect, under certain conditions, in the parvocellular pathway and by the anatomy of some retinal circuits. Contrarily to this idea, we will demonstrate that chromatic opponency is mainly a property of the spatial multiplexing of cone receptors (Hérault, 1996; Alleysson, 1999). We will explain how the midget cells' response may exhibit chromatic opponency without any special circuit and how the segregation between luminance and chrominance should be operated in the visual cortex (Billock, 1995; Kingdom and Mullen, 1995). The special case of S cones with their specific circuitry will be explained.

5.1. Color Coding in the Retina

We will not describe the nature of color and color measurements here. The interested reader will find comprehensive developments of this topic in many excellent books (Le Grand, 1970; Boynton, 1979; Wyszecki and Stiles, 1982; Mollon, 1983; De Valois and De Valois, 1990; Kaiser *et al.*, 1996; Fairchild, 1998). Instead, we will emphasize on the modeling of what happens to color signals throughout the visual pathway.

In fact, some neuro-anatomical data strongly induce the idea that color is processed at the retinal level: two types of horizontal cells are apparently

141

linked to color: the H1 cells which principally connect L and M cones and sporadically S cones, and the H2 cells always connected to S cones and also to L and M cones (Kolb *et al.*, 1980; Boycott *et al.*, 1987; Dacey *et al.*, 1996). Moreover, there exists a specific circuit linked to S cones.

However, we will explain in the subsequent sections how the responses of the bipolar cells may lead to the belief of a special circuit for chromatic opposition, though not necessary, and how the segregation between luminance and chrominance may (and should) be done principally at the cortical level (Billock, 1995; Kingdom and Mullen, 1995). Let us review some experimental facts that support this proposal.

- According to recent results, if horizontal cells were responsible of the chromatic opposition, the receptor fields of the RED-GREEN cells would be 20 times wider than actually observed (Croner and Kaplan, 1995; Lee *et al.*, 1998).
- It is certain that, in human and non-human primates, horizontal cells have an achromatic response, contrarily to other species.
- There is no specific circuit for color in Midget cells (Calkins and Sterling, 1996) and no morphological distinctions have been made between the ganglion cells facing the L and M cones. There is, however, an exception: a special circuit exists for S cones (Dacey and Lee, 1994; Calkins *et al.*, 1998; Smithson and Mollon, 2004), which leads to "Blue-ON / Yellow-OFF" cells, however no "blue-OFF / Yellow-ON" cells have been observed in intracellular records (Martin, 1998). We will explain why this special circuit is necessary.

5.1.1. *Biological Data*

The three types of cones are identified by their spectral sensitivity curves (Smith and Pokorny, 1975; Schnapf *et al.*, 1988; Jacobs, 1996) : L (orange-red), M (yellow-green) et S (blue). It should be noted that there is a significant overlap in the spectral sensitivities of L and M cones whereas the S cones sensitivity is shifted more towards short wave-lengths (see Figure 5.1). We will broach this topic and analyze its functional consequence later.

Cones are disposed on a centered hexagonal grid, the mesh of which is more or less regular, on the retinal surface (Marc and Sperling, 1977;

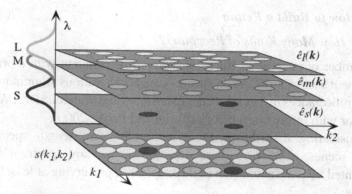

Figure 5.1. Spatio-chromatic sampling of the cone mosaic. 1 — There are three kinds of photoreceptors to sample the wavelength spectrum, 2 — The photoreceptors are spatially multiplexed: there is one photoreceptor per location on the retinal surface.

Roorda and Williams, 1998), leading to a 15% optimization of the space covering with respect to a square grid.

This is true in the fovea but some perturbation of this scheme appears with increasing eccentricity because of the presence of rods. Because we have only one photoreceptor per spatial position, the three cone types are spatially multiplexed (Martinez-Uriegas, 1990).

This kind of spatio-chromatic sampling with a limited number of receptor types results from a trade-off between the need of a good spatial accuracy and a sufficient representation of the spectral shape of objects: with one kind of receptor, the spatial accuracy would be maximum, but no color representation would be possible, whereas with many receptor types tuned to many spectral bandwidths to precisely represent the objects' chromatic spectra, they would be so scattered over the retinal surface that only a weak spatial accuracy would be obtained (Atick *et al.*, 1990; Torralba *et al.*, 1998).

Let us recall that photoreceptors are coupled by Gap-junctions, a mean to reduce the structural noise due to the variability in receptors' thresholds and gains. Under some hypothesis, it is supposed that this coupling would concern only receptors of the same type L, M or S (this seems rather logical). But it has not yet been proved in primates. Even if this hypothesis is not fully confirmed, we will see that it may be more accurate for color decoding, though not fully necessary: there is an overlap in spectral sensitivities. Why would there not be an overlap in the spatial domain?

5.1.2. *How to Build a Retina*

5.1.2.1. *How Many Kinds of Receptors?*

The number of cone types (three) appears to be coherent with a principal component analysis (PCA) of spectra carried on various illuminants and objects reflectances in natural scenes (Marimont and Wandel, 1992; Wandel, 1995), or with information theory (Turiel and Parga, 2000).

Considering the statistical distribution of wave-length spectra for natural scenes, a PCA on these data shows that any spectrum can be represented by a weighted sum of three spectral, preserving at least 90% of variance:

- the first curve (first principal component) is all positive and represents the mean spectrum of all possible reflected lights, its shape is very similar to the Human visibility function V_λ,
- the other curves (second and third principal components) are positive and negative, resembling more or less the chromatic oppositions Blue/Yellow and Red/Green observed in psychophysics.

There being three curves is thus compatible with the existence of three kinds of cone photoreceptors. However, let us recall that, due to biochemical constraints, the photoreceptors' output signals can be only positive. The receptors' sensitivities cannot directly map the curves of the second and third principal components. It is thus necessary to combine the three receptors' outputs to find out these positive/negative components. Because the photoreceptors' sensitivities are given by the nature of photopigments, the mapping may not be perfect.

Under these conditions, it seems that three kinds of photoreceptors are sufficient to represent natural light spectra. They would not be prohibitive to preserve spatial accuracy.

5.1.2.2. *Choice of the Spatial Distribution of Spectral Sensitivities*

We have three receptor types. Could there be a procedure to distribute them spatially in some optimal fashion? One idea would be to distribute them uniformly on a centered hexagonal grid, which is compatible with the number of three, providing the same regular sampling for each photoreceptor.

However, this scheme is not the one the Nature has chosen. In fact, if we look at the spatial details available according to wave-lengths in natural

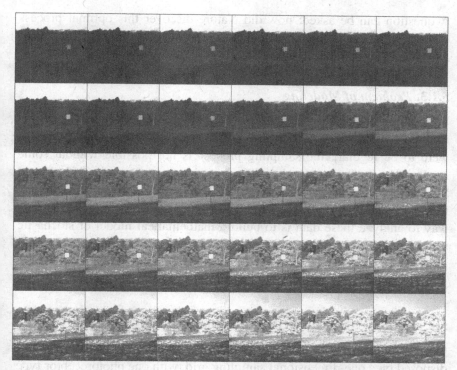

Figure 5.2. A color image of a natural scene viewed through 30 wave-length windows of 10 nanometers, from 350 to 650 nanometers. See that the higher spatial accuracy lies in longer wave-lengths. Redesigned after the data from Bristol Hyperspectral Images Database (see Parraga *et al.*, 1998).

scenes, we see that the longer wave-lengths (in the range of green to red) carry much more details than the shortest wave-lengths (in the range of violet to green). This fact has been experimentally observed with multispectral photo cameras (Parraga, 1995; Brelstaff *et al.*, 1995; Parraga *et al.*, 2002) as shown at Figure 5.2.

Consequently, in order to preserve spatial accuracy, the spatial density of long wave-length receptors should be much higher than that of short wave-length receptors. In other words, there is no need to densely sample an image in the range of blue light (recall Shannon's theorem on sampling). Furthermore, if the spectral sensitivities of long wave-length receptors significantly overlap, their sum will be equivalent to a "super photoreceptor" with an optimal spatial sampling rate in the range of green to red lights. It is exactly the scheme of sampling that we find on the surface of the retina.

A question can be asked: how did Nature discover this optimal process without previously knowing Shannon's theorem and Principal Component Analysis? This remains a mystery...

5.1.3. *Problem of Modeling*

We are now facing the problem that the photoreceptors' sampling grid is covered by non-identical densities of the three receptor types, even if it were to be regular,: the sampling process is not possible without some randomness. In fact, anatomical data show that the proportions of L, M and S cones in the retina vary from 1 to 10% for S cones and that the ratio of L to M cones varies from 1/3 to 3 in human primates. Consequently, the only way to analyze these data is to build a mathematical model of stochastic sampling. Presenting this model directly would seem very complex. We prefer to begin with a simplified regular model which will exhibit the main principles and properties of spatial mutiplexing.

5.2. A Simple Model in Order to Understand

Our simplified model considers three photoreceptors types, regularly disposed on a one-dimensional sampling grid, with one photoreceptor type at each spatial position. Even with this simplification, this model will be able to reveal that the property of chromatic oppositions is inherent to the spatial multiplexing, and that the retina can process color without any special circuitry. This model will also reveal that the parvocellular pathway transmits information about both luminance and chrominance. The extension to a regular sampling on a two-dimensional centered hexagonal grid will be shown.

5.2.1. *One-Dimensional Regular Sampling for Three Receptors*

5.2.1.1. *Basic Equations and Frequency Representation*

Let us consider the foveal and parafoveal areas of the retina. Our color receptors (cones) are regularly spaced on a one-dimensional grid of mesh Δl, which will be taken as a unity in the sequel: $\Delta l = 1$. The model does not allow the presence of rods. The three kinds of receptors — we will name these Red, Green and Blue — are supposed to be equally numerous (1/3 each) and uniformly scattered:

$$- R - G - B - R - G - B - R - G - \cdots$$

Let us now imagine a *fictive* retina with *three* color receptors *at every* sampling step k, giving the following signals: $r(k), g(k), b(k),$ $r(k + 1), g(k + 1), \ldots$ The frequency spectrum of a signal sampled every Δl is known to be periodic with periodicity $1/\Delta l = 1$. Similarly, the frequency spectrum of a signal sampled every $3\Delta l$ is periodic with periodicity $1/3\Delta l = 1/3$ (Figure 5.3). The "real" retina appears as the subsampling (every $3\Delta l$) of these three signals, each color sample being shifted by one sample from the preceding one. The functions, which allow to transform the fictive retina signals into the one of the real retina are the following modulation functions. Respectively, for red, green and blue samples, they are:

$$r'(k) = r(k)\frac{1}{3}\left[1 + 2\cos\left(2\pi\frac{k}{3}\right)\right] \neq 0 \qquad \text{for } k = 3p$$

$$g'(k) = g(k)\frac{1}{3}\left[1 + 2\cos\left(2\pi\frac{k-1}{3}\right)\right] \neq 0 \quad \text{for } k = 3p + 1 \qquad (5.1)$$

$$b'(k) = b(k)\frac{1}{3}\left[1 + 2\cos\left(2\pi\frac{k+1}{3}\right)\right] \neq 0 \quad \text{for } k = 3p + 2.$$

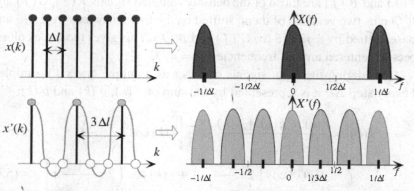

Figure 5.3. Basics of the photoreceptors' multiplexing. Left: signal domain; top: the receptors' sampling positions $x(k)$; bottom: the sampling of one cone type, every third position $x'(k)$; the generating function is shown in light gray. Right: representation in the frequency domain $X(f)$; top: for the sampling positions; bottom: for the one-every third sampling $X'(f)$.

It is now interesting to consider the corresponding signals in the frequency spectrum domain:

$$R'(f) = R(f) * \frac{1}{3}\left[\delta(f) + \delta\left(f - \frac{1}{3}\right) + \delta\left(f + \frac{1}{3}\right)\right]$$

$$G'(f) = G(f) * \frac{1}{3}\left[\left\{\delta(f) + \delta\left(f - \frac{1}{3}\right) + \delta\left(f + \frac{1}{3}\right)\right\} \cdot e^{-j2\pi f}\right]$$

$$B'(f) = B(f) * \frac{1}{3}\left[\left\{\delta(f) + \delta\left(f - \frac{1}{3}\right) + \delta\left(f + \frac{1}{3}\right)\right\} \cdot e^{+j2\pi f}\right],$$

$$(5.2)$$

or, taking into account the property of the delta distribution:

$$R'(f) = R(f) * \frac{1}{3}\left[\delta(f) + \delta\left(f - \frac{1}{3}\right) + \delta\left(f + \frac{1}{3}\right)\right]$$

$$G'(f) = G(f) * \frac{1}{3}\left[\delta(f) + \delta\left(f - \frac{1}{3}\right) e^{-j2\pi/3} + \delta\left(f + \frac{1}{3}\right) e^{+j2\pi/3}\right]$$

$$B'(f) = B(f) * \frac{1}{3}\left[\delta(f) + \delta\left(f - \frac{1}{3}\right) e^{+j2\pi/3} + \delta\left(f + \frac{1}{3}\right) e^{-j2\pi/3}\right].$$

$$(5.3)$$

This means that the frequency spectra of the sub-sampled signals $R'(f)$, $G'(f)$ and $B'(f)$ are those of the densely sampled signals $R(f)$, $G(f)$ and $B(f)$ plus two versions of them, shifted on the frequency axis by $\pm 1/3$ and phase-shifted by $\pm j2\pi/3$ for $G'(f)$ and $B'(f)$. This gives the three colors' spectra, centered around frequencies $f = n/3$.

Considering the input signals of cones as a unique one $s(k)$, sampled at every step Δl: it is represented by the sum of $r'(k)$, $g'(k)$ and $b'(k)$:

$$s(k) = \frac{r(k) + g(k) + b(k)}{3} + \frac{2}{3}\left[r(k)\cos\left(2\pi\frac{k}{3}\right)\right.$$

$$\left. + g(k)\cos\left(2\pi\frac{k-1}{3}\right) + r(k)\cos\left(2\pi\frac{k+1}{3}\right)\right]. \quad (5.4)$$

The first term of $s(k)$ represents the luminance signal $l(k)$ as the sum of the three components, sampled at every step Δl. Its frequency spectrum is hence with a period $f = 1$ (see Figure 5.4). The second term represents

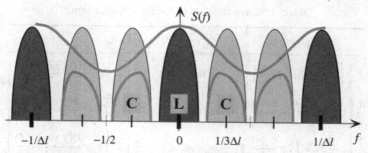

Figure 5.4. Spatial frequency representation of the Luminance and chrominance signals. In dashed lines: the low-pass filter for cones coupled regardless of their spectral type and the resulting effect on chrominance (see text).

the chrominance information. Its frequency spectrum is centered every $f = n \pm 1/3$. It is a high-frequency signal looking like the modulation of a carrier at frequency 1/3:

$$S(f) = \frac{1}{3}[R(f) + G(f) + B(f)]$$
$$+ \frac{1}{3}\left[R\left(f - \frac{1}{3}\right) + G\left(f - \frac{1}{3}\right)e^{+j2\pi/3} + B\left(f - \frac{1}{3}\right)e^{-j2\pi/3}\right]$$
$$+ \frac{1}{3}\left[R\left(f + \frac{1}{3}\right) + G\left(f + \frac{1}{3}\right)e^{-j2\pi/3} + B\left(f + \frac{1}{3}\right)e^{+j2\pi/3}\right].$$

$$(5.5)$$

Consequently, it already appears that the luminance signal can be extracted from $s(k)$ by a simple spatial frequency low-pass filtering. Conversely, the chrominance signal can be isolated by a high-pass filtering.

We can now perform the demodulation of the chrominance signal by multiplying it by each of the modulation functions and by low-pass filtering, which provides the results given in Table 5.1.

The result of this demodulation (right column of Table 5.1) exhibits an effect of color oppositions: Red-Cyan, Green-Magenta and Blue-Yellow. This effect is well known in psychophysics and in visual neurophysiology (Bushbaum and Gottschalk, 1983; Chichilnisky and Wandel, 1999), though in reality it concerns only Red-Green and Blue-Yellow colors. Notice that the three demodulations are not independent. The third one gives the

Table 5.1. Demodulation of the chrominance signal.

Multiplying by		Gives
$\dfrac{1}{3}\left[1 + 2\cos\left(2\pi\dfrac{k}{3}\right)\right]$	in front of R	$\dfrac{2}{3}\left[R(f) - \dfrac{G(f) + B(f)}{2}\right]$
$\dfrac{1}{3}\left[1 + 2\cos\left(2\pi\dfrac{k-1}{3}\right)\right]$	in front of G	$\dfrac{2}{3}\left[G(f) - \dfrac{R(f) + B(f)}{2}\right]$
$\dfrac{1}{3}\left[1 + 2\cos\left(2\pi\dfrac{k+1}{3}\right)\right]$	in front of B	$\dfrac{2}{3}\left[B(f) - \dfrac{R(f) + G(f)}{2}\right]$

Blue-Yellow opposition and the difference between the first two ones gives the Red-Green opposition.

5.2.1.2. *What Happens in the Cones Layer*

Under the hypothesis of cones coupled only within the same type, the cone circuit is equivalent to three independent and non-overlapping low-pass filters, the output of which are multiplexed the same way. So there is no consequence on color: the chromatic organization is not affected by the coupling of cones.

If the coupling is made independently of the cone types (Hsu *et al.*, 2000), the overall spectrum is low-pass filtered: luminance is weakly affected, but there is some noticeable modification in the spectral shapes of chrominance (dashed lines in Figure 5.4). The result after demodulation will give a slight high-pass effect on chromatic opposition signals.

5.2.1.3. *What Happens in the Outer Plexiform Layer*

The transfer function of the Outer Plexiform Layer is known to be spatio-temporally high-pass. Because it removes the low frequencies, this layer affects only the luminance signal (Figure 5.5), leaving the chrominance at the output of the cone layer unchanged.

This model is also valid for the Midget cells of the Parvocellular pathway: this explains why a same Parvocellular cell in the Lateral Geniculate Nucleus exhibits a response to both luminance and chrominance, the former being high-pass filtered, the latter being slightly low-pass filtered. Furthermore, because in our retinal model the Magnocellular pathway is

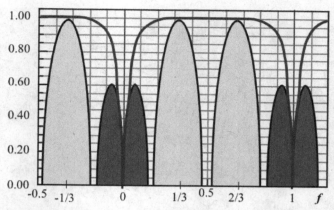

Figure 5.5. Spatial frequency representation of the luminance and chrominance signals in the Outer Plexiform Layer. Chrominance is weakly affected whereas luminance is strongly submitted to the high-pass filtering of this layer.

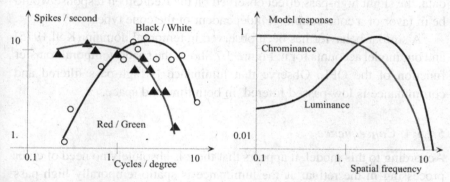

Figure 5.6. Spatial frequency spectra for luminance and chrominance. Left: data in LGNd of Monkey, redrawn after von Blanckensee's data (1981). Right: data obtained from the model.

issued from a low-pass purely spatial filtering from the Inner Plexiform Layer, this readily explains why the Magnocellular pathway does not convey any chromatic information.

The properties we have derived from this model compare favorably with biological data obtained in Parvocellular cells of the cat's LGN reported by De Valois and De Valois in 1990 (Figure 5.6) or from other data (Granger and Hurtley, 1973). Supposing that there is no measurement error in the recorded

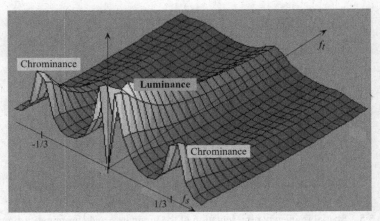

Figure 5.7.   Spatio-temporal transfer function of the OPL and its effect on luminance and chrominance signals.

data, the slight high-pass effect observed on the Red/Green response would be in favor of a cone coupling, independent of the cone types.

A similar behavior has been observed in temporal domain (Kell,1975) and our model accounts for it: Figure 5.7 Shows the spatio-temporal transfer function of the OPL. Observe that luminance is high-pass filtered and chrominance is low-passed filtered, in both time and space.

5.2.1.4. *Consequence*

According to this model, it appears that there is absolutely no need of color processing in the retina: as the luminance is spatio-temporally high-pass filtered, the chromatic information goes through the retinal circuitry almost without any perturbation. If this is true for the biological retina (and it seems to be), it has an important consequence on the economy of neural matter in the optic nerve part, corresponding to the fovea: instead of transmitting three signals (luminance plus two chromatic oppositions) for each receptor, there is only one Midget cell conveying both luminance and chrominance in a multiplexed scheme with its neighbors. This represents a 66% savings of neural matter.

But here, a question may be asked: if it is not in the retina, where does the decoding of color lie in the visual system? And precisely where does the chromatic opposition appear? According to our model of extraction of luminance and chrominance, spatial filtering and demodulation being simple operations (weighted summations and products), they are likely to take place in the primary visual cortex where chromatic opponent cells have been observed in the blobs of cytochrome-oxydase, the physiology of which begins to be well understood (Johnson *et al.*, 2001; Conway *et al.*, 2002; Beaudot and Mullen, 2005).

5.2.2. *The Case of Two Dimensional Regular Sampling*

In the two-dimensional case of regular sampling, an optimal solution for three receptor types is the one of a centered hexagonal grid. In this case, the photoreceptors are in equal density as shown below:

$$
\begin{array}{c}
\textsf{B R G B R G} \\
\textsf{R G B R G B R} \\
\textsf{B R G B R G} \\
\textsf{R G B R G B R} \\
\textsf{B R G B R G}
\end{array}
$$

The preceding equations result in a form which is a bit more complicated than that of the form in the one-dimensional linear case. We will not show them here. They may be retrieved in many computer vision lectures or in the doctoral dissertation of David Alleysson (1999). The result of the corresponding spatial transfer function of the retina is given in Figure 5.8. It can be seen that the Fourier transform of an hexagonal mesh is also an hexagonal mesh in the frequency domain.

The luminance is centered in the low spatial frequencies and high-pass filtered (a shape of a volcano at the center of the image). The chrominance signal is disposed around these luminance centers at the summits of hexagons. Despite the beautiful aspect of this representation, no new property emerges with respect to the one-dimensional case.

5.3. Model of Random Sampling

Let us come back to the model of Figure 5.1 and let us suppose that the three kinds of receptors L, M, et S are randomly distributed on the retinal surface according to a centered hexagonal grid (Roorda and Williams, 1999).

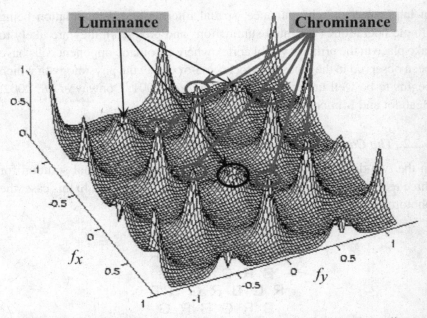

Figure 5.8. Spatial transfer function of the retina with a regular hexagonal sampling and the locus for luminance and chrominance in the frequency domain.

There is only one receptor type at each location (they are spatially multiplexed) and their respective probabilities to be present at a location are: p_l, p_m and p_s. Of course, we have: $p_l + p_m + p_s = 1$. These probabilities are different between species and between individuals within a same species. The model may take any value, but we will choose experimental values as proposed in (De Valois and De Valois, 1993), that is $p_l = 10/16$, $p_m = 5/16$, and $p_s = 1/16$, knowing that according to authors, the ratio between L and M cones may vary from $1/3$ to 3 and that p_s may be lower than 1%.

We will now present a mathematical model for this type of sampling in order to find its properties.

5.3.1. *Sampling Function for One Receptor*

In order to simplify the expressions, we will study a one-dimensional model of signals. The case of two dimensions would give the same properties. Let

$C_i(x)$, $i = l$, m, or s, three fictive color components continuous signals in space, which are sampled by corresponding random functions: $\hat{e}_i(x)$: $C_i(x) \Rightarrow \hat{C}_i(x) = C_i(x) \cdot \hat{e}_i(x)$. The sampling functions $\hat{e}_i(x)$ are Dirac impulses (distributions) at positions $k\,\Delta l = k_i\,\Delta l$:

$$\hat{e}_i(x) = \sum_{k \in k_i} \delta(x - k\,\Delta l)$$

The probability of finding an impulse at position $k\,\Delta l$ is p_i, the probability of finding no impulse at this position is $1 - p_i$ or $\sum_{j \neq i} p_j$.

The random sampling function $\hat{e}_i(x)$ may be represented as the sum of a regular sampling like a "Dirac Comb" $\delta_{\Delta l}(x)$ with the weight p_i, and a zero-mean random function $\tilde{e}_i(x)$ with positive values $(1 - p_i)$ and negative values $(-p_i)$, as indicated Figure 5.9:

$$\hat{e}_i(x) = p_i\,\delta_{\Delta l}(x) + \tilde{e}_i(x) \tag{5.6}$$

Each color receptor signal comprises two components:

- the first one is regularly sampled with a spatial mesh Δl,
- the second one modulates a zero-mean random function.

This last point reminds the technique of modulation on a random carrier used in high-security communications: the frequency spectrum is the convolution of the random carrier spectrum by the one of the modulating signal. The result is a random signal, the power of which is distributed on a wide frequency band. The modulation signal can be recovered if the random

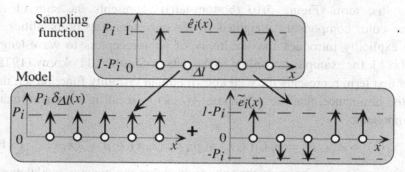

Figure 5.9. Random sampling function $\hat{e}(x)$ and its model seen as the sum of a regular sampling $\delta_{\Delta\lambda}(x)$ h amplitude P_i and a zero-mean random function $\tilde{e}(x)$.

carrier is known. This technique of modulation is known to be very resistant to parasitic noise.

5.3.2. Model of the Random Multiplexing for Three Receptors

5.3.2.1. Principle

Finally, the global sampled signal is the sum of the three sampled color components $\hat{s}(x) = \sum_i C_i(x)\, \hat{e}_i(x)$, or expliciting $\hat{e}_i(x)$:

$$\hat{s}(x) = \sum_i p_i\, C_i(x)\, \delta_{\Delta l}(x) + \sum_i C_i(x)\, \tilde{e}_i(x). \tag{5.7}$$

In order to avoid expressions with Dirac distributions, we will now express the signals under their discrete form, in function of the variable k, taking a unit sampling step ($\Delta l = 1$). The resulting sampled signal $s(k)$ is the sum of the three color channels $C_i(k)$, sampled by the random functions $\hat{e}_i(k)$:

$$s(k) = C_l(k) \cdot \hat{e}_l(k) + C_m(k) \cdot \hat{e}_m(k) + C_s(k) \cdot \hat{e}_s(k). \tag{5.8}$$

5.3.2.2. Towards the Luminance/Chrominance Representation

By replacing the functions $\hat{e}_i(k)$ by their decomposition into a regular sampling plus a zero-mean random function, we get:

$$s(k) = \left\{ \sum_{i \in l,m,s} p_i \cdot C_i(k) \right\} + \left\{ \sum_{i \in l,m,s} C_i(k) \cdot \tilde{e}_i(k) \right\} \tag{5.9}$$

The first term (Figure 5.10 (bottom-left)) represents the sum of the three color components, weighted by the corresponding probabilities. If we explicitly introduce the sensitivity of photoreceptors to wave-length $C_i(\lambda, k)$, for example, according to the data of Smith and Pokorny (1975), this first term represents the well known human visibility function V_λ, that is the luminance, that we will name $A(\lambda, k)$ to recall it is an achromatic component:

$$V_\lambda(k) = A(\lambda, k) = p_l\, C_l(\lambda, k) + p_m\, C_m(\lambda, k) + p_s\, C_s(\lambda, k). \tag{5.10}$$

The second term represents the sum of the color components modulations. We will name it chrominance. Notice an important fact: the appearance of

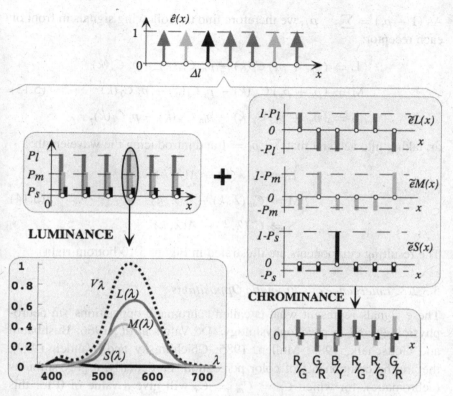

Figure 5.10. Model of chromatic random sampling: luminance appears at each sampling step; chrominance exhibits chromatic oppositions.

chromatic oppositions. If we multiply this term by one of the three random sampling functions, for example $\hat{e}_i(k)$, we obtain:

$$\sum_{j \in l,m,s} C_j(k) \cdot \tilde{e}_j(k) \cdot \hat{e}_i(k). \qquad (5.11)$$

This operation corresponds to the observation of the signal in front of receptor number i. In this sum, the term for $j = i$ gives $(1 - p_i)C_i(k)$ and, as $\hat{e}_i(k)$ and $\hat{e}_j(k)$ cannot have simultaneously the same value 1. The other terms, when $j \neq i$, give $\left(-p_j C_j(k)\right)$, that is:

$$\hat{e}_i(k) \sum_{j \in l,m,s} C_j(k)\, \tilde{e}_j(k) = (1 - p_i)\, C_i(k) - \sum_{j \neq i} p_j\, C_j(k), \qquad (5.12)$$

As $(1 - p_i) = \sum_{j \neq i} p_j$, we therefore find the following signals in front of each receptor:

$$L \Rightarrow (p_m + p_s)\, C_l(k) - p_m\, C_m(k) - p_s\, C_s(k)$$

$$M \Rightarrow (p_l + p_s)\, C_m(k) - p_l\, C_l(k) - p_s\, C_s(k) \qquad (5.13)$$

$$S \Rightarrow (p_m + p_l)\, C_s(k) - p_m\, C_m(k) - p_l\, C_l(k),$$

or, taking into account that $\sum_i p_i = 1$ and introducing the wavelength:

$$L \Rightarrow C_l(\lambda, k) - A(\lambda, k)$$

$$M \Rightarrow C_m(\lambda, k) - A(\lambda, k) \qquad (5.14)$$

$$S \Rightarrow C_s(\lambda, k) - A(\lambda, k)$$

The resulting components are illustrated in Figure 5.10(bottom-right).

5.3.2.3. *Emergence of Chromatic Oppositions*

These signals represent what is called "chromatic oppositions" in neuro-physiology and in psychophysiology (De Valois *et al.*, 1966; Bushbaum and Gottschalk, 1983; Mullen, 1985; Chichilnisky and Wandel, 1999), the are the components of color perception. Particularly, a grey stimulus (achromatic), for which $C_l = C_m = C_s$ will give a value of 0 for this second term. Observe that for the S photoreceptor, the chromatic opposition signal is not directly weighted by p_s, which provides it with the same relative importance than the others, even if the S cones are in small number! Following the data of De Valois (1993), for which $p_l = 10/16$, $p_m = 5/16$ and $p_s = 1/16$, and introducing the wave-length variable, we obtain the following result:

$$L \Rightarrow [6\, C_l(k, \lambda) - 5\, C_m(k, \lambda) - 1 C_s(k, \lambda)]/16$$

$$M \Rightarrow [11\, C_m(k, \lambda) - 10\, C_l(k, \lambda) - 1\, C_s(k, \lambda)]/16 \qquad (5.15)$$

$$S \Rightarrow [15\, C_s(k, \lambda) - 5\, C_m(k, \lambda) - 10\, C_l(k, \lambda)]/16.$$

5.3.3. *Comparison with Biological Data*

We have seen that our model exhibits two main properties of color perception: the human visibility curve $V(\lambda)$ and the chromatic oppositions.

It is worth noticing that the achromatic part $A(\lambda, k)$ is available at every sampling step (no sub-sampling for luminance) and that the chromatic part lies in the high frequencies domain for it results from zero-mean functions. The shape of $A(\lambda, k)$, compared to the curve $V(\lambda)$) for an individual, makes it possible to determine his photoreceptors' ratios, the wave-length sensitivity of photopigments being constant among individuals.

Concerning the chromatic part, we have three signals, the wave-length sensitivities of which are interesting to draw (Figure 5.11a). We have, facing the L cones a mainly "L − M" (Red-Green) component, facing the M cones a mainly "M − L" (Green-Red) component and facing the S cones a mainly "S − (L + M)" (Blue-Yellow) component.

This last signal strikingly compares to the Blue/Yellow opposition component of Jameson and Hurvich (1955) as shown in Figure 5.11b. Considering that we had formerly three signals (L, M, S) and that our model

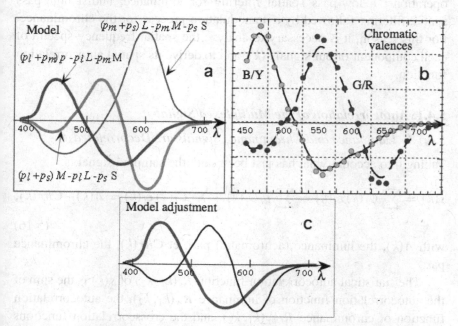

Figure 5.11. Chrominance. (a) variations of the chromatic components of the model with wavelength, (b) psychophysical evaluation of the opponent channels R − G and B − Y (after Jameson and Hurvich, 1955, with kind permission), see the similarity between the first curve of the model and the B/Y opponent channel. (c) the combination of second and third curves in (a) which maps closely the shape of the G/R opponent channel.

transforms them into four linear combinations (one luminance component and three chrominance components), we can replace two of these by one, or by some linear combination of them. Doing this for a particular ratio makes it possible to closely map the Red/Green opposition of Jameson and Hurvich (figure 5.11c).

5.4. Frequency Spectrum of the Random Sampling

Our mathematical model of random sampling highlights signals which correspond to perceptual components. However, these signals are not explicit in the retina. How would it be possible to extract these components, and at which level? This is the question we will answer in this section. We will show that the separation between luminance and chrominance (first and second terms of equation 5.0) can be done by simple filtering operations: a low-pass spatial filtering for luminance, and a high-pass spatial filtering followed by a demodulation process for the chrominance. For this reason, it is necessary to analyze the spatial frequency spectra of each component of our signal $s(k)$, and to derive its spatial autocorrelation function.

5.4.1. *Autocorrelation of the Multiplexed Signal*

5.4.1.1. *Luminance and Chrominance Signals are Decorrelated*

Taking into account what has just been said, the sampled signal is:

$$s(k) = \sum_i C_i(k)\,\hat{e}_i(k) = \sum_i p_i C_i(k) + \sum_i C_i(k)\,\tilde{e}_i(k) = A(k) + Chr(k),$$

$$(5.16)$$

with $A(k)$, the luminance (achromatic) part, et $Chr(k)$, the chrominance part.

The statistical autocorrelation function $R_s(k_1, k_2)$ of $s(k)$ is the sum of the autocorrelation function of luminance $R_A(k_1, k_2)$, the autocorrelation function of chrominance $R_{Chr}(k_1, k_2)$, and the crosscorrelation functions between luminance and chrominance $R_{A,Chr}(k_1, k_2)$ et $R_{Chr,A}(k_1, k_2)$:

$$R_s(k_1, k_2) = R_A(k_1, k_2) + R_{Chr}(k_1, k_2) + R_{A,Chr}(k_1, k_2) + R_{Chr,A}(k_1, k_2)$$

$$(5.17)$$

Let us first calculate the third and fourth terms. The third one is:

$$R_{A,Chr}(k_1, k_2) = E\left[\sum_i p_i\, C_i(k_1) \cdot \sum_j C_j(k_2) \cdot \tilde{e}_j(k_2)\right], \quad (5.18)$$

or, because of the statistical independence between the color components $C_i(k)$ and the random sampling functions $\tilde{e}_i(k)$:

$$R_{A,Chr}(k_1, k_2) = \sum_{i,j} p_i\, E\left[C_i(k_1) \cdot C_j(k_2)\right] \cdot E\left[\tilde{e}_j(k_2)\right]. \quad (5.19)$$

As $E[\tilde{e}_j(k)]$ is zero, by definition, the crosscorrelation between luminance and chrominance is zero (these two signals are therefore mutually orthogonal). This is also valuable for the fourth term of equation (5.17). The statistical autocorrelation function of $s(k)$ is:

$$R_s(k_1, k_2) = R_A(k_1, k_2) + R_{Chr}(k_1, k_2) \quad (5.20)$$

5.4.1.2. *Autocorrelation of Luminance*

Let us calculate now the autocorrelation function of luminance $R_A(k_1, k_2)$:

$$R_A(k_1, k_2) = E\left[\sum_i p_i C_i(k_1) \cdot \sum_j p_j C_j(k_2)\right]. \quad (5.21)$$

This results readily in: $R_A(k) = \sum_{i,j} p_i p_j R_{Cij}(k)$, the signals being supposed stationary up to the second order, at least in an observation window of the size of the fovea. $R_{Cij}(k)$ is the crosscorrelation between the sampled color components $C_i(k)$ and $C_j(k)$. Separating the cases when $i = j$ and when $i \neq j$, we obtain the final expression for the autocorrelation of luminance (the achromatic component):

$$R_A(k) = \sum_i p_i^2 R_{Ci}(k) + \sum_{i,j \neq i} p_i p_j R_{Cij}(k), \quad (5.22)$$

$R_{Ci}(k)$ is the autocorrelation of a color component $C_i(k)$.

5.4.1.3. *Autocorrelation of Chrominance*

Let us now calculate the autocorrelation of chrominance $R_{Chr}(k_1, k_2)$. It will be derived from those of the luminance $R_A(k_1, k_2)$ and of the global signal $R_s(k_1, k_2)$. We have, for the global signal:

$$R_s(k_1, k_2) = E\left[\sum_i C_i(k_1)\,\hat{e}_i(k_1) \cdot \sum_j C_j(k_2) \cdot \hat{e}_j(k_2)\right], \qquad (5.23)$$

and, supposing that the signals and their sampling functions are stationary at order two, and that they are independent:

$$R_s(k) = \sum_{i,j} R_{Cij}(k)\, E\left[\hat{e}_i(k_1)\,\hat{e}_j(k_1 - k)\right]. \qquad (5.24)$$

We now have to separately process the cases when $i = j$ and when $i \neq j$.

For i = j, we have:

$$E[\hat{e}_i(k_1)\,\hat{e}_j(k_1 - k)] = \Pr[\hat{e}_i(k_1) = 1 \cap \hat{e}_i(k_1 - k) = 1],$$

with:

$$\Pr[\hat{e}_i(k_1) = 1 \cap \hat{e}_i(k_1 - k) = 1]$$
$$= \Pr[\hat{e}_i(k_1) = 1] \times \Pr[\hat{e}_i(k_1 - k) = 1 \mid \hat{e}_i(k_1) = 1],$$

that is, for a Markovian process (see Section 5.4.1.4):

$$E[\hat{e}_i(k_1)\,\hat{e}_i(k_1 - k)] = p_i\left[p_i + (1 - p_i)\left(\frac{b - p_i}{1 - p_i}\right)^{|k|}\right], \qquad (5.25)$$

where b is the first diagonal term of the transition matrix of the Markov process. The case of a renewal process would correspond to $b = p_i$. Figure 5.12 gives the autocorrelation functions for three cases: $b > p_i$, $b = p_i$ and $b < p_i$. The first case means that samples of the same kind are likely to agglomerate to each other. The second one is purely random (like a Poisson process). The last one means that if a receptor is present at one place, it has a lower chance to be present at the neighboring place.

We notice, in the Markovian case, a damped oscillation around frequency $1/2$, which reveals a tendency to alternate between "presence/absence" for a same receptor type. When k increases, this rule vanishes and the presence follows the rule of chance.

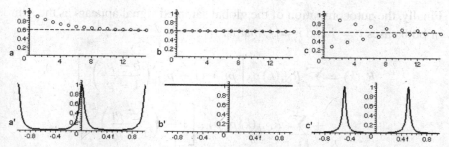

Figure 5.12. Top: Autocorrelations of sampling functions. a) b > p, b) b = p, c) b < p. Bottom: the corresponding frequency spectra. The only interesting case is c − c': it means that a receptor of some type is likely to be absent if the same type is present nearby. The corresponding spectrum is a modulation of a random carrier around the Nyquist frequency.

For $i \neq j$, we have:

$$E[\hat{e}_i(k_1)\,\hat{e}_j(k_1 - k)] = \Pr[\hat{e}_i(k_1) = 1 \cap \hat{e}_j(k_1 - k) = 1],$$

with:

$$
\begin{aligned}
&\Pr\left[\hat{e}_i(k_1) = 1 \cap \hat{e}_j(k_1 - k) = 1\right] \\
&= \Pr\left[\hat{e}_i(k_1) = 1\right] \cdot \Pr\left[\hat{e}_i(k_1 - k) = 0 \mid \hat{e}_i(k_1) = 1\right] \\
&\quad \times \Pr\left[\hat{e}_j(k_1 - k) = 1 \mid \hat{e}_i(k_1 - k) = 0\right].
\end{aligned}
$$

The third factor is the probability for $\hat{e}_j$ to be present when $\hat{e}_i$ is absent: $p_j / \sum_{j \neq i} p_j = /(1 - p_i)$. Then we have, for the same Markovian process:

$$E[\hat{e}_i(k_1)\hat{e}_j(k_1 - k)]$$

$$= p_i \left\{ 1 - \left[p_i + (1 - p_i)\left(\frac{b - p_i}{1 - p_i}\right)^{|k|} \right] \right\} \frac{p_j}{1 - p_i}, \qquad (5.26)$$

that is, after simplification:

$$E\left[\hat{e}_i(k_1)\,\hat{e}_j(k_1 - k)\right] = p_i\,p_j \cdot \left[1 - \left(\frac{b - p_i}{1 - p_i}\right)^{|k|} \right]. \qquad (5.27)$$

Finally, the autocorrelation of the global sampled signal appears as the sum of both cases $i = j$ and $i \neq j$, that is:

$$R_s(k) = \sum_i R_{Ci}(k) p_i \left[p_i + (1 - p_i) \left(\frac{b - p_i}{1 - p_i} \right)^k \right]$$

$$+ \sum_{i, j \neq i} R_{Cij}(k) p_i \, p_j \cdot \left[1 - \left(\frac{b - p_i}{1 - p_i} \right)^{|k|} \right]$$

and, re-arranging:

$$R_s(k) = \sum_{i,j} p_i \, p_j R_{Cij}(k) + \sum_i p_i \left(\left(\sum_{j \neq i} p_j \right) R_{Ci}(k) - \sum_{j \neq i} p_j R_{Cij}(k) \right)$$

$$\times \left(\frac{b - p_i}{1 - p_i} \right)^{|k|} \tag{5.28}$$

which is to be compared to equation (5.20): $R_s(k) = \{R_A(k)\} + \{R_{Chr}(k)\}$.

We recognize in the first term the autocorrelation function of luminance $R_A(k)$ which is a combination of the auto- and cross-correlations of the color components.

In the second term, we find the autocorrelation of chrominance $R_{Chr}(k)$ which appears as a combination of the products of the chromatic opposition autocorrelation functions, by exponentials modulating a periodic function of period 2 samples. In fact, if $b < p_i$, we have:

$$\left(\frac{b - p_i}{1 - p_i} \right)^{|k|} = \cos \left(2\pi \frac{k}{2} \right) e^{-|k| \ln(\alpha_i)}, \quad \text{with } \alpha_i = \left(\frac{b - p_i}{1 - p_i} \right).$$

We will now assess the power frequency spectrum of $s(k)$ in order to show how luminance and chrominance information can be separated. But before moving on, let us explain the Markovian process for the reader who may be interested.

5.4.1.4. *Markovian process*

This process is defined by the fact that the probability $b = p(1, n+1|1, n)$ to have a photoreceptor of type i at position $n+1$ depends only on the existence of the same receptor at position n. We can write the global probabilities as:

$$p(1, n+1) = p(1, n+1 \mid 1, n) p(1, n) + p(1, n+1 \mid 0, n) p(0, n)$$
$$p(0, n+1) = p(0, n+1 \mid 1, n) p(1, n) + p(0, n+1 \mid 0, n) p(0, n),$$

or, in a vector-matrix form:

$$\begin{bmatrix} p(1, n+1) \\ p(0, n+1) \end{bmatrix} = \begin{bmatrix} p(1, n+1 \mid 1, n) & p(1, n+1 \mid 0, n) \\ p(0, n+1 \mid 1, n) & p(0, n+1 \mid 0, n) \end{bmatrix} \cdot \begin{bmatrix} p(1, n) \\ p(0, n) \end{bmatrix},$$

(5.29)

knowing that $p(1, n) = p, \forall n$ and $p(0, n) = 1 - p, \forall n$ and that if $p(1, n+1 \mid 1, n) = b$, then $p(0, n+1 \mid 1, n) = 1 - b$, we have:

$$\begin{bmatrix} p(1, n+1) \\ p(0, n+1) \end{bmatrix} = \begin{bmatrix} b & \frac{p(1-b)}{1-p} \\ 1-b & 1 - \frac{p(1-b)}{1-p} \end{bmatrix} \cdot \begin{bmatrix} p(1, n) \\ p(0, n) \end{bmatrix},$$

(5.30)

This expression defines the probabilities for the first neighbor. If we want the probabilities for the kth neighbor, we write:

$$\begin{bmatrix} p(1, n+k) \\ p(0, n+k) \end{bmatrix} = \begin{bmatrix} b & \frac{p(1-b)}{1-p} \\ 1-b & 1 - \frac{p(1-b)}{1-p} \end{bmatrix}^k \cdot \begin{bmatrix} p(1, n) \\ p(0, n) \end{bmatrix} = \mathbf{T} \cdot \begin{bmatrix} p(1, n) \\ p(0, n) \end{bmatrix}.$$

In this case, the probability to have one sample of type i at position $n + k$, knowing that there is one sample of type i at position n, is given by the first diagonal term of matrix $\mathbf{T}$:

$$p(1, n+k \mid 1, n) = p + (1 - p) \left(\frac{b - p}{1 - p} \right)^k.$$

(5.31)

This is the formula that has been used for Figure 5.12.

5.4.2. *Spectral Representation*

In order to estimate our signal in the spatial frequency domain, let us estimate its spectral power density function $S_s(f)$ by taking the Fourier transform of its autocorrelation function:

$$S_s(f) = \left\{ \sum_{i,j} p_i \, p_j \, S_{Cij}(f) \right\}$$

$$+ \left\{ \sum_i p_i \left((1 - p_i) \, S_{Ci}(f) - \sum_{j \neq i} p_j \, S_{Cij}(f) \right) \right\}$$

$$\times \frac{1}{2} \left\{ \left[\delta \left(f - \frac{1}{2} \right) + \delta \left(f + \frac{1}{2} \right) \right] \times \frac{1}{1 + (f/\ln(\alpha_i))^2} \right\}.$$

$$(5.32)$$

the first term $\sum_{i,j} p_i \, p_j \, S_{Cij}(f)$ represents the power spectral density of luminance (classically in $1/f^2$ for images). It is centered around frequency 0. The second term represents the power spectral density of chrominance: the chromatic oppositions $(1 - p_i) \, S_{Ci}(f) - \sum_{j \neq i} p_j S_{Cij}(f)$, modulating a carrier $[\delta(f - 1/2) + \delta(f + 1/2)]$ centered around spatial frequency $f = \pm 1/2$), with a slight spectral spreading $1/1 + (f/\ln(\alpha_i))^2$ because of the stochastic character of the sampling process (second convolution product). Figure 5.13 shows the spatial frequency spectrum of the multiplexed color signal. We will see in our next section how this fact can be exploited to separate luminance and chrominance information.

To be more accurate, calculations should take into consideration the spatiochromatic correlations between color components. The results would be of the same order, in the case of much more complicated developments. For more information, see David Alleysson's PhD thesis (Alleysson, 1999).

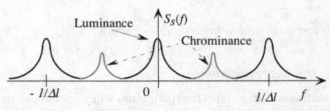

Figure 5.13. Power spectral density of the sampled signal according to a Markov process.

5.4.3. *Decoding Luminance and Chrominance*

5.4.3.1. *Spatial Filtering and Demodulation*

The spectral shapes of luminance and chrominance allow the extraction of each of these terms by a simple filtering process in the spatial domain.

Luminance will be extracted from a spatial low-pass filtering as indicated Figure 5.14. Notice that this filter is centered on frequencies $f = n/\Delta l$, meaning that its output is distributed on the fine sampling mesh (Δl), which allows a luminance image with the best spatial resolution, despite the multiplexing receptors of different types. Its frequency spectrum is:

$$S_A(f) = \left\{ \sum_{i,j} p_i \, p_j \, S_{Cij}(f) \right\}, \qquad (5.33)$$

that is, the sum of the frequency spectra of each component, plus the interspectra between components.

Chrominance will be extracted first, according to the following procedure:

- apply a complementary (high-pass) spatial filter.
- demodulate the resulting signal, that is, simply multiply by $\cos(2\pi k/2)$ to center its spectrum around frequency 0 (bottom of Figure 5.15),

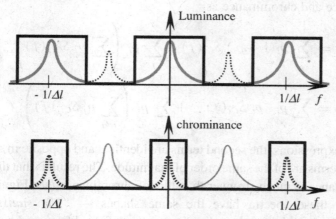

Figure 5.14. Low-pass filtering to extract luminance (Top), demodulation, and low-pass filtering to extract chrominance (bottom).

- apply a simple low-pass filter which will allow to isolate the chromatic oppositions on the finer sampling mesh Δl.

After demodulation, the spectrum of chrominance is:

$$S_{Chr}(f) = \left\{ \sum_i p_i \left((1 - p_i) S_{Ci}(f) - \sum_{j \neq i} p_j S_{Cij}(f) \right) \right\}$$
$$\times \frac{1}{4} \left\{ \frac{1}{1 + (f/\ln(\alpha_i))^2} \right\} \tag{5.34}$$

Observations.

- The recovery of chrominance will not be perfect because its spectrum is widened by the convolution product. In fact, the demodulation of chrominance components should be done by multiplication by the sampling functions $\hat{e}_i(k)$ themselves.
- The spectra of luminance and modulated chrominance may partially overlap. This could generate aliasing in high spatial frequencies, as it has already been observed in biology (Osorio *et al.*, 1998).

5.4.3.2. *Luminance — Chrominance Energy Ratio*

From equations (5.33) and (5.34), we can express the power spectra of luminance and chrominance as:

$$Chr \Rightarrow \sum_i p_i(1 - p_i) S_{Ci}(f) - \sum_i p_i \left(\sum_{j \neq i} p_j S_{Cij}(f) \right)$$
$$Ach \Rightarrow \sum_i p_i \ \ p_i S_{Ci}(f) \ \ + \sum_i p_i \left(\sum_{j \neq i} p_j S_{Cij}(f) \right). \tag{5.35}$$

In both expressions, the second terms are identical and opposite in sign, and the first terms are of the same order of magnitude. The result is that the power of luminance is much stronger than that of chrominance (see Figure 5.15). Because these spectra have the same shapes — *"The statistical 1/f amplitude spatial-frequency distribution is confirmed for a variety of chromatic conditions across the visible spectrum"* (Paraga *et al.*, 1998) — the

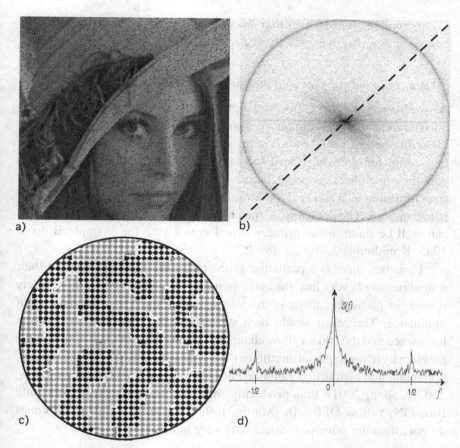

Figure 5.15. Example of random color sampling: a) input image (see the cover of this book), b) its frequency spectrum, c) an example of the sampling function, d) the frequency spectrum profile along the dashed line of b). This random sampling is given for two kinds of receptors. Note that the chrominance spectrum lies on the circle of radius $1/2$ and that its power is lower than the one of luminance.

spectrum of luminance is much wider than that of chrominance. This explains:

- why the biological and psychophysical observations reveal a wider frequency band for luminance than for chromatic oppositions channels,
- why physical measurements of light ask some questions, *"Our analysis suggests that natural scenes are relatively rich in high-spatial-frequency*

chrominance information that does not appear to be transmitted by the human visual system. . ." (Paraga *et al.*, 1998).

5.4.3.3. *Economy of Connections, the case of S Cones*

As we have said before, this model proves that the coding of cone signals into luminance and chrominance signals does not need to be processed in the retina by special circuits, simply because it is an inherent property of the spatial multiplexing (be it regular or random). This model complies with the principle of economy, because instead of transmitting three signals (one for luminance and two for chrominance), only one multiplexed signal is transmitted. The decoding of the chromatic and achromatic components can well be done in the primary visual cortex (see for example Billock, 1995; Kingdom and Mullen, 1995).

However, there is a particular problem for the S cones: because their spatial density is very low, the corresponding frequency spectrum is widely spread, in particular, towards the low spatial frequencies, the domain of luminance. The result would be a strong aliasing phenomenon between luminance and the Blue/Yellow channel. It seems that Nature has solved the problem by using a special circuit for the S cones only involving Db6 bipolar, horizontal and bistratified ganglion cells (Dacey and Packer, 2003; Gouras, 2003; Calkin, 2001), thus producing the Blue/Yellow channel with only Blue-ON/Yellow-OFF cells (Martin, 1998). The small number of S cones do not affect the principle of economy very much.

5.5. Application to CCD Cameras

The spatial multiplexing we have in the retina is currently used in low-cost photo cameras. In these cameras (mono CCD), there is one unique sensor, the pixels of which are masked by color filters, mostly of the type of the Bayer Color Filter Array (Bayer, 1976), according to the following scheme:

R G R G R G R G
G B G B G B G B
R G R G R G R G
G B G B G B G B

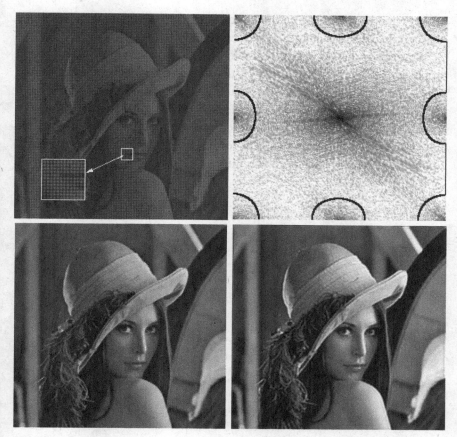

Figure 5.16. Example of a mono-CCD camera image. Top left: with its Bayer CFA. Top right: the resulting 2D frequency spectrum with the boarders of the low- and high-pass filters. Bottom left: the luminance recovered by low-pass filtering (with the chrominance/luminance aliasing in insert). Bottom right: the three-colors per pixel image obtained after demodulation. See colored figure in Color Plate i.

An example of this sampling array is shown in the insert at Figure 5.16, top-left. The first technique to recover a three-colors per pixel image has been to interpolate each of the three sub-sampled images, and to combine them again. This gives a poor result: it produces a very blurred image unsuitable for usual quality, unless starting from a very dense sampling around 2000 × 2000 pixels. During the last decade, a number of optimized techniques have been proposed (Adams, 1998; Glotzbach *et al.*, 2001;

Trussell and Hartwig, 2002), often requiring a large amount of calculations to ascertain a correct quality of the result.

Starting from our model of color retina, a new technique has been proposed, based on the structure of the frequency spectrum of the sensor signal, and with an adequate demodulation process, just as we have seen above (Alleysson *et al.*, 2005). This technique has the advantage of a very low computational charge for an acceptable accuracy. Here again, we see that the use of a biological model results in a savings of computational complexity, hence of hardware circuitry.

Chapter 6

Non-Linear, Irregular and Non-Stationary Processes

In this chapter, we will see how linear processes are insufficient to extract information in a signal. Irregular sampling avoids aliasing effects (moiré) in images. Non-linear transfer functions may allow to extract interesting variables from multiplicative structures. Global and local adaptive processes may circumvent the effects of contexts.

6.1. Photoreceptor's Irregular Spatial Sampling

In this section we will see how the spatial sampling of the retina by the midget ganglion cells induces, through some topological transform, an equivalent sampling of the visual scene with highly interesting properties on an ecological point of view. It provides direct accesses to:

- an invariant description of images whatever the reading distance is,
- a measure of the time-to-contact in case of ego-motion,
- a measure of the perspective information around the point of fixation.

6.1.1. *Biological Data*

The first exhaustive and accurate analysis of the anatomy of retinal cells in humans is due to Kolb *et al.* (1992), see an interesting review in (Silveira *et al.*, 2005). The topography of photoreceptors has been widely studied (Osterberg, 1935; Williams, 1988; Curcio *et al.*, 1990; Jonas *et al.*, 1992). At the retinal output, the anatomy and topography of ganglion cells has been studied in primates (Long and Fisher, 1983; Perry and Cowey, 1985), in humans (Hebel and Holländer, 1983; Rodieck *et al.*, 1985; Curcio and Allen, 1990; Dacey and Petersen, 1992), in humans, macaques, cats and marmossets, by Goodchild *et al.* (1996), see a review in (Martin and Grüner, 2003). More recent studies can be found in Brett *et al.* (2005) and Szmajda *et al.* (2005).

All the experimental data show a strong decrease of the cone photoreceptors density, from fovea to periphery, with a marked intriguing rebound around 80–85° of eccentricity (Figure 6.1a). In order to know if this rebound may be of some use in the sampling of the visual field, we need to procede to a more in-depth study of the retinal sampling.

That is, we first need to take into account the spatial distribution of the ganglion cells at the retinal output (see Figure 6.1b) and combine it with the topological transform of the visual field through the optical geometry of the eye.

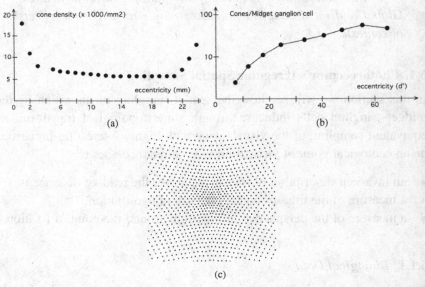

(a)

(b)

(c)

Figure 6.1. Retinal sampling. (a) density of cones versus retinal eccentricity in mm (redrawn from combined data, from Curcio *et al.*, 1990 and Osterbeg, 1935). (b) number of cones per midget ganglion cell, versus retinal eccentricity in degrees (redrawn from Goodchild *et al.*, 1996 data). (c) example of a corresponding model of spatial sampling, on the basis of a centered hexagonal sampling process.

6.1.2. Geometry of the Eye

The seminal work on eye optics and geometry was done by Listing in 1845. One and a half century later, new and more complex models were developed. See for example (Rabbetts and Bennett, 1998; Atchison and Smith, 2000; Westheimer, 2006). For our case, we will take the simplified but accurate model of Drasdo and Fowler (1974).

This model assumes the retina is a sphere with center C and radius $R = 11.06$ mm (Figure 6.2). The fovea is situated in such a way that the visual axis presents an offset of $\alpha_{e0} = -5°$ with respect to the optical axis (axis of symmetry of the eye). According to the data reported by Drasdo and Fowler (1974) for the optical system of the eye (cornea+lens), if a ray of light issued from a point of coordinates (X,Z) and passing through N makes an external angle α_e with respect to the visual axis ($\alpha_e' - \alpha_{e0}'$ with respect to the optical axis), it is refracted at the secondary nodal point N' with an internal angle $\alpha_i' = 0.8\alpha_e'$. It hits the retina at a distance x from the fovea such that $x = \beta r$, with $\beta = \beta' - \beta_0$. The distance between N' and C is $a = 4.6$ mm. The geometry results in: $\tan(\alpha_i) = \tan(\alpha_i' - \alpha_{i0})$,

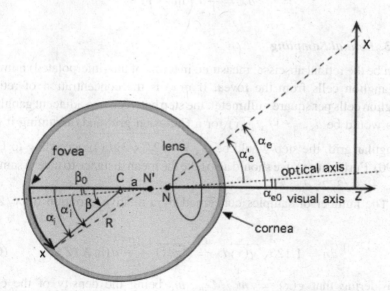

Figure 6.2. Simplified schema of the geometry of the eye, designed according to the data of Drasdo and Fowlers' model (1974).

with

$$\tan(\alpha_i') = \frac{R\sin(\beta')}{R\cos(\beta') + a} \quad \text{and} \quad \tan(\alpha_{i0}) = \frac{R\sin(\beta_0)}{R\cos(\beta_0) + a}.$$

Let us calculate the relative variation of X/Z, that is, the logarithmic derivative of $\tan(\alpha_e)$ with respect to X:

$$d\left(\ln\frac{X}{Z}\right) = \frac{d\tan(\alpha_e)}{\tan(\alpha_e)} = \frac{2}{\sin(2\alpha_e)}\frac{d\alpha_e}{d\alpha_i}\frac{d\alpha_i}{d\beta}\frac{d\beta}{dx}dx,$$

or

$$d\left(\ln\frac{X}{Z}\right) = \frac{2.5}{\sin(2.5\alpha_i)}\frac{1 + \frac{a}{R}\cos(\beta - \beta_0)}{1 + \left(\frac{a}{R}\right)^2 + 2\frac{a}{R}\cos(\beta - \beta_0)}\frac{1}{R}dx \doteq Ldx. \quad (6.1)$$

In this expression, α_i should be replaced by its value as a function of $\beta = x/R$ and β_0 should be replaced by its value as a function of α_{e0}. Thus, the variation of the retinal abscissa x with respect to the relative variation of X/Z is:

$$dx = \frac{1}{L}d\left(\ln\frac{X}{Z}\right).$$

6.1.3. *Retinal Sampling*

Let n be the retinal abscissa, measured in terms of the (interpolated) number of ganglion cells from the fovea. If $g(x)$ is the concentration of retinal ganglion cells per square millimeter, the step between two adjacent ganglion cells would be $d_{sq} = 1/\sqrt{g(x)}$ for a Cartesian grid and, assuming it is a triangular grid, the step would be $d_{tri} = \sqrt{2/(\sqrt{3}g(x))}$ according to Hild (1990). Furthermore, we should also take the mean distance to meet a sample $d_{mean} = 3\sqrt{3}/2\pi\, d_{tri}$.

The number of samples concerned by a relative variation of X/Z is thus:

$$dn = 1.125\sqrt{g(x)}dx = \sqrt{g(x)}\frac{1.125}{L}d(\ln X/Z). \quad (6.2)$$

Considering that $g(x) = m_c/M_{cpg}$, m_c being the density of the cone photoreceptors and M_{cpg} being the number of photoreceptors per midget ganglion cells, the curve for the density $g(x)$ is the ratio of the two functions given in Figures 6.1a and 6.1b, provided that the eccentricity in degrees is converted into mm of retinal abscissa. The experimental data for M_{cpg} are interpolated by the following formula:

$$M_{cpg} = 2 + 48 \frac{x^2 + x^3 + x}{x^2 + x^3 + x + 700}.$$

Let us compare, on the same graph, the distribution $g(x)$ and the curve $1.125/L$ (Figure 6.3).

Figure 6.3 shows the linear spatial density $g(x)$ of the retinal output samples versus eccentricity (mm) for biological data. The continuous curve represents simultaneously the variation of function L. We clearly see that the ratio $\sqrt{g(x)}/L$ will be almost constant over all eccentricities. Figure 6.4 verifies this assumption: for the two data sets available, the ratio $n_0 = 1.125\sqrt{g(x)}/L$ remains roughly constant for all eccentricities, except for the foveal region.

Given this, we can say that the number of retinal samples concerned with a relative variation of coordinates is constant whatever the eccentricity may be.

$$dn = n_0 d\left(\ln \frac{X}{Z}\right) = n_0\left(\frac{dX}{X} - \frac{dZ}{Z}\right), \quad \text{with } n_0; \ 60\,\text{mm}^{-1}.$$

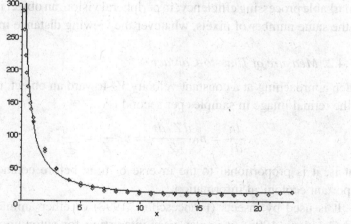

Figure 6.3. Dots: Linear spatial density $g(x)$ of retinal samples versus eccentricity (mm) from the data of Curcio *et al.* (1990) and Osterberg (1935). Continuous curve: function 6L, suitably scaled for comparison.

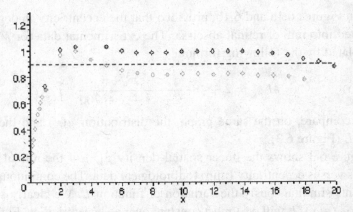

Figure 6.4. Number of retinal samples per unit of relative scene coordinates variation for both data of Osterberg (lower curve) and Curcio (upper curve). This number is almost constant over a wide range of eccentricities.

6.1.4. Consequences: Advantages of Logarithmic Coordinates

6.1.4.1. Constancy of Peripheral Description

When reading a newspaper, a word or a letter of size ΔX at distance X from the fixation point is analyzed by a constant number of samples $\Delta n = n_0 \Delta X / X$, whatever the reading distance is. This property implies a remarkable processing efficiency: in peripheral vision, an object is described by the same number of pixels, whatever the viewing distance may be.

6.1.4.2. Measure of Time-to-Contact

When approaching at a constant velocity V_Z toward an object, the velocity of the retinal image in samples per second is:

$$\frac{dn}{dt} = n_0 \frac{dZ/dt}{Z} = n_0 \frac{V_Z}{Z} = \frac{n_0}{T_c},$$

that is, it is proportional to the inverse of time before contact, a highly important ecological information.

It is used by insects (Franceschini, 1998) or other animals (Laurent and Gabbiani, 1998), it is also very interesting for autonomous robotics (Franceschini *et al.*, 1992; Bolduc and Levine, 1998; Viollet and Franceschini, 1999).

This fact is well known in computer vision for robotics. See Rojers and Schwartz (1990), Tistarelli and Sandini (1993), and more recently Adams and Horton (2003) and Calow *et al.* (2005). It should be noted that for a static scene, this velocity is independent of the viewing eccentricity, except at the fovea.

6.1.4.3. *Interest for Retinal Filters Design*

Many studies on the variation of sensitivity to contrast as a function of spatial frequency have already been carried on. Virsu and Näsänen (1978) have plotted the corresponding curves for various retinal eccentricities (Figure 6.5a), showing that the retinal image is more and more low-pass filtered when eccentricity increases. However, they observed that, if the frequency on the graph is normalized to the ganglion cells sampling frequency at each eccentricity, all the curves superimpose (Figure 6.5b)! This means that the spatial filters of the retina, which are discrete filters, have exactly the same coefficients, whatever the eccentricity may be. This contributes to a very economic coding scheme. Galvin *et al.* (1997) showed experimentally that subjects systematically judged that a peripheral stimulus was less blurred than it actually was. They called this effect "peripheral sharpness overconstancy". This may be due to the high-pass behavior of the retinal filter.

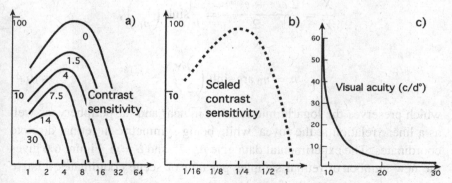

Figure 6.5. (a) contrast sensitivity in the human subject as a function of retinal eccentricities 0, 1.5, 4, 7.5, 14, 30 d° (from Virsu & Näsänen, 1978); (b) contrast sensitivity with frequency, scaled by the inverse of ganglion cells spacing; (c) visual acuity versus eccentricity.

It should be noticed that the retinal acuity (Anderson and Hess, 1990; Sere *et al.*, 2000) deduced from these curves (Figure 6.5c) exactly follows our curve of retinal sampling $g(x)$ of Figure 6.3. As the ganglion cells directly map to the primary visual cortex, this property is related to the "cortical magnification factor" (Cowey and Rolls, 1974; Wässle *et al.*, 1990; Duncan and Boynton, 2003), as we will see later.

6.1.5. Consequences at the Cortical Level

From the expression $dn = n_0 d(\ln(X/Z))$ we can deduce by a simple integration the spatial transform between the scene coordinates and the sample number:

$$n = n_0 \ln \frac{X}{Z} + n_0 \ln b, \qquad (6.3)$$

b being a constant to be determined. However, this formula does not hold, neither at the fovea where the logarithm tends toward $-\infty$, nor for negative values of n where the logarithm is not defined.

In order to overcome these problems, let us rewrite the preceding formula as:

$$\frac{X}{Z} = \frac{1}{b} e^{\frac{n}{n_0}}.$$

And, in order to symmetrize, let us write:

$$\frac{X}{Z} = \frac{2}{b} \frac{e^{\frac{n}{n_0}} - e^{-\frac{n}{n_0}}}{2} = \frac{2}{b} \sinh\left(\frac{n}{n_0}\right),$$

or

$$n = n_0 \arg \sinh\left(\frac{b}{2}\frac{X}{Z}\right), \qquad (6.4)$$

which preserves the logarithmic relation in near and far periphery, as well as a linear relation at the fovea, while being symmetric in retinal discrete coordinates. The experimental data give n_0; 75 and $b = 4$. Figure 6.6 gives the new number of retinal samples per unit of scene coordinates relative variation, according to the formula (6.4).

Let us now see what happens when viewing the perspective of a scene. Figure 6.7 shows the problem in one dimension for more simplicity: a line

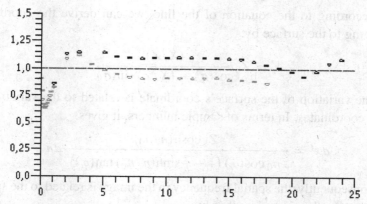

Figure 6.6. Same as Figure 6.4, but with $n = n_0 \arg \sinh(bX/2Z)$. Observe that the constancy of the ratio (Number of retinal samples per unit of relative scene coordinates variation) is more accurately respected over all eccentricities.

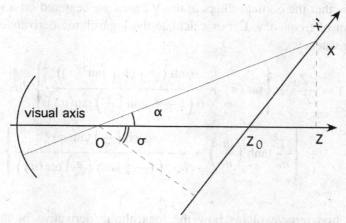

Figure 6.7. Simple geometry of a scene perspective. O: center of vision and world coordinates, Z_0: distance to the point of fixation, α: visual eccentricity, σ: angle of slant of the object's surface, x': coordinate belonging to the surface.

lying on a surface and passing through the point of fixation at a distance Z_0 is described by its equation in scene coordinates:

$$Z = X \tan(\sigma) + Z_0, \tag{6.5}$$

σ being its angle of slant with respect to the observer.

According to the equation of the line, we can derive the coordinate belonging to the surface by:

$$x' = Z_0 \frac{(X/Z)}{\cos(\sigma) - (X/Z)\sin(\sigma)}.$$

The variation of the surface's coordinate is related to the variation of world coordinates. In terms of sample numbers, it gives:

$$dx' = \frac{Z_0 \cosh(n/n_0)}{\frac{b}{2}n_0 \cos(\sigma)\left(1 - \frac{2}{b}\sinh(n/n_0)\tan(\sigma)\right)^2}dn.$$

Consequently, the spatial frequency in the image is related to the spatial frequency on the viewed surface by:

$$f_i = \frac{Z_0 \cosh(n/n_0)}{\frac{b}{2}n_0 \cos(\sigma)\left(1 - \frac{2}{b}\sinh(n/n_0)\tan(\sigma)\right)^2}f_s. \qquad (6.6)$$

We know that the cortical filters in the V1 area are centered on a scale of logarithm of frequency. Let us calculate the logarithmic derivative of the image frequency:

$$d\ln(f_i) = \frac{dZ_0}{Z_0} + \left\{\tan(\sigma) + \frac{\sinh\left(\frac{n}{n_0}\right)(1 + \tan^2(\sigma))}{b\left(1 - \frac{2}{b}\sinh\left(\frac{n}{n_0}\right)\tan(\sigma)\right)}\right\}d\sigma$$

$$+ \left\{\frac{1}{n_0}\tanh\left(\frac{n}{n_0}\right) + \frac{\cosh\left(\frac{n}{n_0}\right)\tan(\sigma)}{bn_0\left(1 - \frac{2}{b}\sinh\left(\frac{n}{n_0}\right)\tan(\sigma)\right)}\right\}dn.$$

$$(6.7)$$

- The first term explains how the logarithmic derivative of the local frequency with respect to time directly gives us the *inverse of time-to-contact* (there is no need to measure any velocity!). This result is independent of the local frequency on the object's surface and of the visual eccentricity, so that it can be integrated over a large region of the image, in order to gain more accuracy.
- The second term shows that the *slant of the surface* can be directly accessed at the fovea when *manipulating* the object by a rotation movement $d\sigma$. That is $\frac{d\ln f_i}{d\sigma} = \tan(\sigma)|_{n:0}$, whatever the object's texture may be.

- The third term, the local frequency gradient in the sampled image, can very simply provide information concerning the slant, at the fovea.

It should be noticed that all these values are fully independent of the local frequency of the surface, provided that there is one (the surface should not be uniform).

6.2. The Random Sampling Model

Despite the apparent regularity of the retinal sampling, there is some randomness in the photoreceptors' positions. The mathematical model of random sampling is not simple and many researchers have worked on this subject in the past decades (Bossomaier *et al.*, 1985; Geisler and Hamilton, 1986) as well as more recently, concerning signal analysis and reconstruction (Aldroubi and Feichtinger, 1998; Petrou *et al.*, 2004; Castano-Moraga *et al.*, 2004). To derive the simplest model of the retinal random sampling, let us start from a regular sampling of step Δx to which we add a random jitter around the positions of the regular sampling such as in Figure 6.8.

For the one-dimensional regular sampling of an image $i(x)$ of step Δx, the model is:

$$\hat{i}(x) = i(x) \cdot \sum_k \delta(x - k\Delta x), \qquad (6.8)$$

and its Fourier spectrum is given by:

$$\hat{I}(f) = \hat{I}(f) * \sum_n \delta(f - n/\Delta x), \qquad (6.9)$$

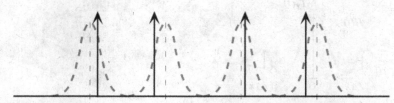

Figure 6.8. Model of 1-D random sampling: samples are distributed around integer positions (dashed lines) with a random jitter, the probability density function of which is drawn as gray bell-shaped curves.

that is, the replication of the image spectrum every $1/\Delta x$ on the frequency axis. In the case of a random jitter, the sampled image is:

$$\hat{i}(x) = i(x) \cdot \sum_k \delta(x - k\Delta x - \xi(k)), \qquad (6.10)$$

where $\xi(k)$ is a random variable with the probability density function $p_\xi(\xi, k)$. The Fourier transform of this sampled image is a random function and cannot be calculated. A simple thing we can do is to estimate the power spectrum of the sampling function:

$$\Gamma_s(f) = \frac{1}{\Delta x^2} |\Pi(f)|^2 \sum_{n=-\infty}^{+\infty} \delta\left(f - \frac{n}{\Delta x}\right) + \frac{1}{\Delta x}(1 - |\Pi(f)|^2, \quad (6.11)$$

where $\Pi(f)$ is the characteristic function (Fourier transform) of $p_\xi(\xi, k)$.

The first term in Equation (6.11) tells us that the replicas of the image spectrum at frequencies $\pm 1/\Delta x$, $\pm 2/\Delta x \ldots$ (of equal amplitudes in the case of a regular sampling) will be attenuated by the function $\Pi(f)$. This is an interesting fact, which implies a substantial reduction of aliasing (moiré effect for images). However the second term indicates that some noise will be added, as experimentally observed (Yellot, 1983). Fortunately, this noise will be negligible in the central region of the spectrum and will concern mainly the high frequencies (see Figure 6.9a).

An example of the 2D spectrum for a limited number of samples and for a square grid is given at Figure 6.9b.

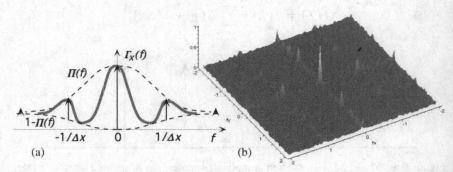

Figure 6.9. Two-dimensional spatial frequency spectrum of the random sampling. Note the strongly attenuated replicas at frequencies $\pm 1/\Delta x$, $\pm 2/\Delta x \ldots$.

To summarize, the random sampling, though adding noise in high frequencies, is essential in order to reduce the aliasing. This is particularly important for the retinal coding of color (as seen in Chapter 5), where the sampling is very sparse, especially for the blue cones.

6.3. Log-Polar Cortical Projections

The first direct evidence of a structured mapping of the visual field onto the cortex has been given by Talbot and Marshall (1941). This was confirmed electrophysiologically 20 years later, and 40 years later, Schwartz (1980) proposed a model as a complex logarithm mapping:

The retinal and cortical coordinates, respectively (x, y) and (X, Y) are represented by complex numbers, $z = x + jy$ and $Z = X + jY$, respectively. The mapping proposed by Schwartz was:

$$Z = \frac{\log[a + z]}{\log[b + z]}, \tag{6.12}$$

a and b corresponding respectively to $0.3°$ and $50°$ of eccentricity. This model fits the biological data fairly well and we will use it in a simplified form.

6.3.1. *Simplified Model*

Let us take the simplified mapping $Z = \log[a + z]$, using either the Cartesian coordinates $z = x + jy$ or the polar coordinates $z = \rho e^{j\theta}$, through a logarithmic transformation, or $Z = \log|a + z| + j \arg(a + z)$.

For some variation of retinal Cartesian coordinates $dz = dx + jdy$ or polar coordinates $dz = e^{j\theta}d\rho + j\rho e^{j\theta}d\rho$, the cortical coordinates variation is:

$$dZ = \frac{dz}{a + z} = \frac{1}{a + z}(dx + jdy) = \frac{z}{a + z}\left(\frac{d\rho}{\rho} + jd\theta\right). \tag{6.13}$$

If this variation is due to a combination of movements: translation ($V_x = dx/dt$, $V_y = dy/dt$), rotation ($V_\theta = d\theta/dt$) and zoom ($V_\rho = d\rho/dt$), we have, in cortical coordinates:

$$\frac{dZ}{dt} = \frac{e^{-jY}}{e^{\text{Re}(Z)}}(V_x + jV_y) + \left(1 - a\frac{e^{-jY}}{e^{\text{Re}(Z)}}\right)\left(\frac{V_\rho}{\rho} + jV_\theta\right). \qquad (6.14)$$

By integrating these variations over a rectangular cortical region $X = [X_0 - \xi, X_0 + \xi]$, $Y = [-Y_1, Y_1]$, we obtain real coefficients for the velocity terms:

$$\int_{X_0-\xi}^{X_0+\xi} \int_{-Y_1}^{Y_1} \frac{dZ}{dt} dX dY$$

$$= \int_{X_0-\xi}^{X_0+\xi} \int_{-Y_1}^{Y_1} \left[\frac{e^{-jY}}{e^X}(V_x + jV_y)\right.$$

$$+ \left.\left(1 - a\frac{e^{-jY}}{e^X}\right)\left(\frac{V_\rho}{\rho} + jV_\theta\right)\right] dX dY$$

$$= e^{-X}4\text{sh}(\xi)\sin(Y_1)(V_x + jV_y)$$

$$+ 4\xi Y_1\left(1 - ae^{-X_0}\frac{\text{sh}(\xi)}{\xi}\frac{\sin(Y_1)}{Y_1}\right)\left(\frac{V_\rho}{\rho} + jV_\theta\right). \qquad (6.15)$$

This will be useful to extract ego-motion information.

6.3.2. *Extraction of Global Motion Information*

The first and second terms of Equation (6.15) contain two different kinds of global information, depending on the cortical region where the integral is performed:

At the locus of projection of the central retina, we have $z = 0$ and $e^{-X} \approx a$. By integrating over a small surface, $\frac{\text{sh}(\xi)}{\xi}\frac{\sin(Y_1)}{Y_1} \approx 1$ and then:

$$\int_{X_0-\xi}^{X_0+\xi} \int_{-Y_1}^{Y_1} \frac{dZ}{dt} dX dY \approx 4\frac{\xi Y_1}{a}(V_x + jV_y), \qquad (6.16)$$

the real and imaginary parts give the retinal *translation* velocities V_x and V_y, as shown in Figure 6.10 b, in the projection region of the fovea.

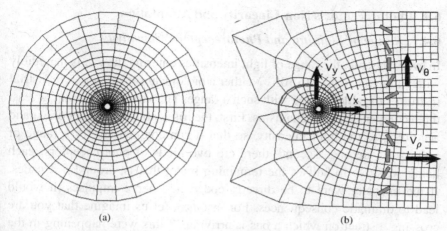

(a)   (b)

Figure 6.10. Log-polar projection. (a) retinal space, (b) cortical space. It is a conformal transformation. The image of a local zone is reproduced with an affinity and a rotation (the small rectangles provide an idea of the type of rotation).

At the region of projection of the retinal periphery, $|z|$ is large and $e^{-X} \approx 0$. We get:

$$\int_{X_0-\xi}^{X_0+\xi} \int_{-Y_1}^{Y_1} \frac{dZ}{dt} dX dY = 4\xi Y_1 \left(\frac{V_\rho}{\rho} + j V_\theta \right), \qquad (6.17)$$

the real and imaginary parts giving respectively the *radial* and *angular* velocities V_ρ and V_θ, as shown Figure 6.10 b, for the region corresponding to the projections of the retinal periphery.

6.3.3. *Conclusion*

The Log-polar model of retino-cortical projection presents an interesting property when considering computational efficiency. The simple estimation of cortical horizontal and vertical mean velocities, allows the estimation of four ego-motion parameters, respectively:

- In the region corresponding to the foveal area, these make it possible to estimate translation motions of the retinal image.
- In the region corresponding to the retinal periphery, these estimate looming and rotation motions.

6.4. Photoreceptor's Non-Linearity and Adaptation

6.4.1. *Range of Lighting and Photoreceptors' Sensitivity*

In every day life, the range of light intensities entering the eye is incredibly wide: it ranges from 1 to 10^{13}. Neither a biological sensor nor an industrial instrument is able to cope with such a range. The visual system has at least three compensation mechanisms. First, the iris is able to contract or expand according to the light intensity, so that it can compensate for a range of around 1.5 decade. Second, there are two kinds of photoreceptors, each being sensitive to half of the remaining scale; that is, around 6 decades. If the intensity were to be directly coded into nerve impulses, it would lead to dramatic consequences. For instance, let us imagine that you are crossing a street on which a bus is arriving. If this were happening in the summer, at noon for example, the light intensity would be maximal and would correspond, let us say to 1000 impulses per second: you would have way enough time to see the bus and take the appropriate decision. If this same situation were to happen in winter, for instance at midnight, let us say without any street lights for example, the light intensity would correspond to 1 impulse for every 20 seconds: you would be absolutely unable to recognize the dangerous situation in time!

Fortunately, in everyday life we never have sudden variations of a 10^6 range of intensities. Most of the time, a maximal range of 1.5 to 2 decades is observed. Our photoreceptors will code for this range by simply adapting to the mean ambient light. Figure 6.11 describes this process. A retinal ganglion cell is recorded (Valeton and van Norren, 1983) for various intensities X of short duration, under different ambient lights X_0, from near dark to 10^6 trolands: we observe that:

- At each ambient level the response curve is S-shaped for logarithmic abscissas.
- This curve translates, following the successive increases of ambient light.

Understanding how these responses are produced has been a challenging task. For more than 25 years since the seminal work of Valeton and van Norren, many studies have been carried out, concerning this modeling procedure (Croner *et al.*, 1993; Beaudot, 1995, 1996; Smirnakis *et al.*, 1997;

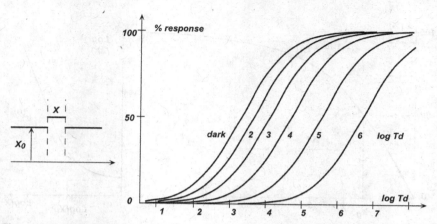

Figure 6.11. Response of a ganglion cell to light pulses X under various conditions of ambient light X_0 (redrawn according to Valeton and van Norren, 1983, with kind permission).

Salinas and Sejnowski, 2001; van Hateren, 2005). The following section will show a simplified model which explains this phenomenon.

6.4.2. *Modeling the Photoreceptor's Non-Linearity*

The mathematical model that is commonly used to describe the S-shapes of Figure 6.11, for the response x to a stimulation X, is the Michaelis-Menten law:

$$x = \frac{X^n}{X^n + X_0^n},\qquad (6.18)$$

n being an exponent between 1 and 2.

X_0 is the value of X that gives 50% of the maximal response. When the exponent is $n = 1$, the Equation (6.18) represents the so-called "Naka–Rushton law" (Rushton, 1965):

$$x = \frac{X}{X + X_0}.\qquad (6.19)$$

It is relatively easy to find the best values of n and X_0 that suitably fit the experimental curves given at Figure 6.11. The model for $n = 1$ is given in linear axes in Figure 6.12a.

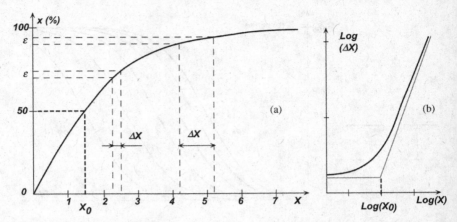

Figure 6.12. Naka–Rushton compression law. (a) in linear coordinates. (b) the corresponding TVI curve in logarithmic coordinates (see text).

However, these coefficients are not very convenient to account for some other data such as the TVI (Threshold Versus Intensity) functions. In biology or in psychophysics, it is difficult to measure an absolute value. Relative values are easier to obtain and produce more reliable results.

The stimulus is held at a given value X and is then given a variation, ΔX: the value of ΔX that produces the just noticeable difference (JND) is considered to be the variation threshold that is able to produce a minimal change of sensation. This is illustrated in Figure 6.12a, where the same "sensation" ε is produced by different values of ΔX, depending on the mean value of X. Figure 6.12b represents the TVI in logarithmic coordinates: the curve presents two asymptotes, one with a slope of 0 and one with a slope of 2.

Unfortunately, most of the experimental data do not confirm the slope of 2. They most often show a slope of 1 and rarely a slope of 2. This result holds as well for biological data on rods and cones (Aguilar and Stiles, 1954; Stiles, 1959) as for psychophysical data (Wyszeki and Stiles, 1982; Spillman and Werner, 1990; Adelson, 1993). An interesting review can be found in (Hahn and Geisler, 1995).

In order to understand this intriguing phenomenon, we will have to consider the biochemical process underlying the light transduction and the way it may adapt to illumination.

6.4.3. *Modeling the Photoreceptor's Adaptivity*

6.4.3.1. *Light Conversion in the Photoreceptor*

The mechanism of light transduction comprises of four steps (McNaughton, 1993). First step: as a consequence of the energy of an incident photon, the visual pigment Rh (Rhodopsin) is isomerized into an active form RH*, which is then free to laterally diffuse within the disc membrane.

Second step: this activated rhodopsin interacts with a number of molecules of transducin (T), which in turn is converted into an active state, thus causing a previously bound molecule of Guanosin diphosphate (GDP) to be exchanged for a triphosphate one (GTP). A high gain results during this operation, because every molecule of Rhodopsin is capable of activating up to 500 molecules of Transducin.

Third step: activated molecules of Transducin switch Phosphodiesterase enzyme molecules (PDE) in a one-to-one fashion into an active state, which is able to catalyze the hydrolysis of $3'5'$-cyclic Guanosin monophosphate (cGMP) into non-cyclic GMP. A further gain factor of 500 occurs.

Final step: the cGMP is known to directly modulate the light-sensitive channels, opening them and allowing the influx of Sodium (Na^+) and Calcium (Ca^{2+}) ions into the photoreceptor.

Because the arrival of a photon causes a decrease in cGMP concentration, the effect of incoming light is to reduce the Na/Ca inward current, and hence to hyperpolarize the photoreceptor membrane. Finally, after the end of illumination, the cGMP concentration returns to its steady state, under the action of the Guanylate cyclase enzyme (Gcy), which catalyzes the synthesis of cGMP from GTP.

Light adaptation occurs at the fourth step, because of a negative feedback mediated by Calcium ions. During the dark steady state, the influx of Ca^{2+} ions is compensated by an equally rapid efflux carried by the active pump exchanging Na^+ with Ca^{2+} and K^+, between the two sides of the cellular membrane. Under the action of light, the light-sensitive channels close and the Ca^{2+} influx decreases. Due to the above-mentioned pump, the internal concentration $[Ca^{2+}]$ decreases.

But the internal Calcium is known to inhibit the activity of Guanylate cyclase. If the $[Ca^{2+}]$ concentration is lowered as a result of light action, the Gcy is less inhibited and then catalyzes the synthesis of cGMP, which in turn will open the light-sensitive channel (Torre *et al.*, 1986). Though not fully proved, it is clear that such a feedback mechanism might be a good candidate to an explanation of light adaptation (Krizaj and Copenhagen, 2002). Because the internal Calcium concentration is the time integral of the net influx, it can act with some time constant in the gain control of light transduction, as it will be shown in the sequel. In order to characterize this process, we will derive a simple mathematical model, inspired from earlier works (Pugh and Lamb, 1993; Hamer and Tyler, 1995; Venkataraman *et al.*, 2003).

6.4.3.2. *Equations for a Simplified Model*

For simplicity's sake, let us consider that the first three steps of light transduction are linear, up to the activation of the PDE enzyme. Then let us examine the enzymatic reactions, which govern the synthesis and hydrolysis of cGMP. To keep things clear, let us name B the cGMP concentration, C the GMP concentration and A the GTP concentration.

These three molecules are embedded in a cyclic reaction as shown Figure 6.13. The component B dissociates into C at a rate β controlled by the incoming light. C dissociates into A at a rate γ under the cell's metabolism control. A dissociates into B at a rate α under the control of the internal Calcium concentration. B opens the ionic channels for Na and Ca, giving rise to the membrane potential Vm (the photoreceptor's output signal), while increasing the internal Calcium concentration through the Ca^{2+} influx.

Let us now write the dynamical equations of dissociation, taking into account that there is no loss of matter, that is, the sum A+B+C is a constant

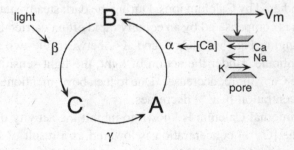

Figure 6.13.   Simplified model of the phototransduction mechanism (see text).

equal to K.

$$\frac{dB}{dt} = -\beta B + \alpha A, \quad \frac{dC}{dt} = \gamma C + \beta B, \quad A + B + C = K. \quad (6.20)$$

Assuming that the chemical reactions are rapid with respect to the membrane time constant, we must just consider the equilibrium state. For the B concentration, we get:

$$B = \frac{\alpha \gamma K}{\alpha \beta + \beta \gamma + \gamma \alpha} \quad (6.21)$$

Considering the normalized output variable $b = \frac{K-B}{K}$, we obtain:

$$b = \frac{\beta}{\beta + \frac{\alpha \gamma}{\alpha + \gamma}},$$

or

$$b = \frac{\beta}{\beta + \beta_0}. \quad (6.22)$$

With this transformation, $b = 0$ when $\beta = 0$ (no light) and $b = 1$ when $\beta = \infty$ (maximum illumination). It is exactly the Naka–Rushton law, with a parameter $\beta_0 = \frac{\alpha \gamma}{\alpha + \gamma}$ subject to a feedback control. By simply choosing the action of Calcium concentration on α of the form $\alpha = \alpha_0 + \frac{\alpha'}{[Ca]}$ and the Calcium concentration as inversely proportional to the mean value of b, we have $\alpha = \alpha_0 + \alpha_1 \cdot b$. For values such that $\alpha_0 \ll \gamma \ll \alpha_1$, the range of β_0 is $\alpha_0 = \beta_{0-\text{inf}} < \beta_0 < \beta_{0-\text{sup}} = \gamma$.

When β is very low, $\beta_0 = \beta_{0-\text{inf}}$ is also low, and when β is very high, β_0 reaches its limit $\beta_0 = \beta_{0-\text{sup}}$, and no more variation is possible. In this last case, the slope of the TVI curve should be 2. At intermediate values of β, because of the negative feedback, the slope should be lower. As illustrated in Figure 6.14, the simulated TVI curve represents the experimental data fairly well, resulting in a slope value 1 at intermediate levels of stimulation, and a slope value 2 near saturation.

6.4.3.3. *Model for Temporal and Spatial Adaptation*

We have said that the Calcium concentration was time-dependent, it being the time integral of the influx. In fact, because of the losses, we should consider it as a low-pass temporal filter of the output signal with a large

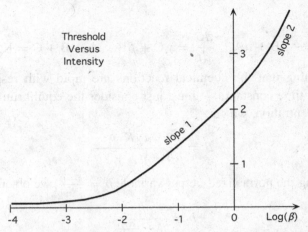

Figure 6.14. Simulated curve of Threshold Versus Intensity for the simple model of light adaptation. Note that slope value 1 is in mid-intensities and slope value 2 is near saturation. Numerical values for this curve are $\alpha_0 = 1$, $\alpha_1 = 50$, $\gamma = 10$.

time constant. In the preceding simulation, we have considered $[C_a]$ as a constant linked to the output signal, thus at its steady state, just as if the receptor would have been submitted to the stimulus for a long time.

If the stimulus is flashed, the internal $[C_a]$ concentration has no time to change, and the TVI curve should present a slope value 2.

Furthermore, the slope of the TVI curve also depends on the background illumination, because, if the stimulus consists of a spot of light with some surrounding illumination, the duration of this surrounding illumination changes the slope of the TVI curve, in the case of electrophysiological experiments (Finkelstein *et al.*, 1988; Walraven *et al.*, 1990) as well as for psychophysical experiments (Adelson, 1993).

This spatial influence on the TVI curve may be explained through the horizontal cells network. The horizontal cells are good candidates because they integrate the surrounding information of a given photoreceptor. If we consider at this point that the feedback synapses of horizontal cells on the photoreceptors should modify the Calcium influx, we have the same influence of the spatial surroundings on the photoreceptor's adaptation as the temporal surroundings, due to the temporal integration of the Ca influx (Figure 6.15).

In summary, the stimulus can either be temporally short or long, and spatially, either narrow or wide. This gives us 4 possibilities for

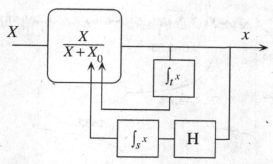

Figure 6.15. Adaptation of ganglion cells to the mean temporal and spatial variations of light intensity.

experimentation, among which, only one would prevent adaptation and lead to a slope value of 2: in the case of a spatially narrow and temporally short stimulus. This could explain why most of the experiments report a slope value of 1.

The model presented here is largely simplified. One can find an extended and remarkable in-depth study, presented by van Hatteren (2005).

6.5. Non-Linearity in the IPL

Photoreceptors are a particular kind of neurons. In fact, all nerve cells present the same aspect of amplitude compression. Whatever the particular mechanism is, we can imagine the same model for all kinds of neurons. Much research has been done on this topic, either electrophysiologically on retinal cells (Freeman, 1991), or in the visual system (Ohzawa, *et al.* 1985). Several theoretical types of research on information transfers have recently been made (Stemmler and Koch, 1999; Yu and Lee, 2005). The case of retinal ganglion cells is of particular interest when dealing with signal pre-processing. It appears to be complementary to the case of photoreceptors.

6.5.1. *Adaptive Non-Linearity in IPL*

The adaptation process in ganglion cells has first been reported by (Shapley and Victor, 1979; Victor, 1987), and later, by Smirnakis *et al.* (1997). It has been reported more recently in the cat's retina (Brown and Masland, 2001), the salamander's retina (Chander *et al.*, 2001; Kim and Rieke, 2001), and the macaque's retina (Kunken *et al.*, 2005).

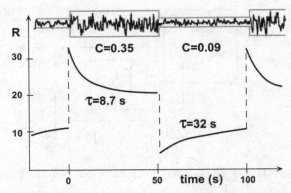

Figure 6.16.   Adaptation of ganglion cells to the mean temporal variations of contrast. Top: flicker stimulus alternating between 0.09 and 0.35 contrast values. Bottom: the mean firing rate of a ganglion cell (redrawn after data from Smirnakis *et al.*, 1997).

The bipolar cells' signal is a high-pass filtered version of the cone signal, and as such, it can be considered as a local estimation of the contrast. As the ON or OFF ganglion cells receive the respectively ON or OFF bipolar cells' signal, their adaptation mechanism applies to the contrast of the retinal image. Figure 6.16 illustrates this aspect: a flicker stimulus is applied to the retina with a contrast alternating between 0.09 and 0.35 every 50 seconds. The simultaneous record of a ganglion cell exhibits an adaptation behavior with a different time constant, according to the sign of the contrast change.

This phenomenon appears separately on both ON and OFF ganglion cells in a similar manner. It has been measured on ganglion cells, but it may also be a property of bipolar cells (Wohrer *et al.*, 2006; Wohrer and Kornprobst, 2008). It was given the name of *Contrast Gain Control* by Shapley and Victor (1979).

6.5.2. *Consequences on the Processing of Images*

6.5.2.1. *OPL: Seeing in Cast Shadows*

The model for the adaptation of photoreceptors is both temporal and spatial, based on a feedback structure:

$$x = \frac{X}{X + X_0}, \quad \text{with} \quad X_0 = X_{00} + x(t, s) \underset{t}{*} h_1(t) \underset{s}{*} h_1(s),$$

that is, the convolution over the time variable by a temporal kernel $h_1(t)$ and the convolution over the (2D) spatial variable s by a spatial kernel $h_2(s)$. In the practical implementation, we will chose a feed-forward structure, which is computationally simpler though not fully equivalent:

$$x = \frac{X}{X + X_0}, \quad \text{with} \quad X_0 = X_{00} + X(t, s) \underset{t}{*} h_1(t) \underset{s}{*} h_1(s). \qquad (6.23)$$

Let us give an example of application with only the spatial adaptation (the temporal adaptation is useful only for video images): the original image Figure 6.17-a represents the entrance of a chapel with a porch under a front roof, on a sunny day. In this case, the range of light is very wide between full sun and shadow. The result for a camera is a dilemma: either it adapts to the shadow and the regions of high luminance will saturate, either it adapts to high luminance and the details in the shadow will be cancelled. By applying Equation (6.23) everywhere in the image, we obtain the image in Figure 6.17b: the gain is increased in regions of low intensities, and decreased in regions of high intensities. The result is a good representation of details in shadows.

However, because the gain is reduced in regions of high intensity, this also results in a reduction of the local contrast, making less details available in these regions.

(a)   (b)

Figure 6.17. Model of photoreceptors adaptation: (a) the original image, (b) the image after the adaptation process: Details can be seen in the shadow.

Figure 6.18.   Model of ganglion cell adaptation: (a) the original image, (b) the image after this second adaptation process applied to the image of Figure 6.17-b. Details now appear both in shadow and in high lights.

6.5.2.2. *IPL: Seeing Details Everywhere*

In order to circumvent this drawback, we use the model of Ganglion cells adaptation in the IPL: the gain will increase in the regions of low contrast. Figure 6.18b shows what happens when this processing is applied to the bipolar cells signal, in the same scene.

6.6. Non-Linearity, Adaptation and Color

6.6.1. *Cone and Luminance Non-Linearities Look the Same*

Let us apply the general non-linearity law: $x = X/(X + X_0)$ for each L, M and S photoreceptor respectively: $l = L/(L + L_0), m = M/(M + M_0)$, and $s = S/(S + S_0)$. Suppose we have a given color of which we vary the intensity. This is modeled by the introduction of a common factor a in the receptor's functions:

$$l = \frac{aL}{aL + L_0}, \quad m = \frac{aM}{aM + M_0} \quad \text{and} \quad s = \frac{aS}{aS + S_0}. \qquad (6.24)$$

We will see how the achromatic component $A = P_l l + P_m m + P_s s$ varies with the common intensity factor a. For simplicity's sake, let us put

$P_l = P_m = P_s s = 1/3$, $x = L/L_0$, $y = M/M_0$ and $z = S/S_0$. We obtain:

$$A = \frac{1}{3} \frac{a\,(x+y+z) + a^2\,2(xy+yz+zw) + a^3xyz}{1 + a(x+y+z) + a^2(xy+yz+zw) + a^3xyz}. \quad (6.25)$$

A priori, this formula does not look like the Naka-Rushton law. However, the behavior of A with respect to a is very similar: observe that for small values of the intensity, the contributions of a^2 and a^3 are near 0 and A follows a Naka-Rushton law with respect to a. When the common factor a increases to higher values, it follows a kind of Michaelis-Menten law with exponent 2 then 3. Figure 6.19-a gives an example of the receptors' responses and of the luminance component A with more realistic percentage values of photoreceptors $P_l = 10/16$, $P_m = 5/16$ and $P_s = 1/16$ for a particular color, already used in Chapter 5.

This is an interesting property because the computational model we have derived in the preceding section is valid not only for the photoreceptors signal, but also for the luminance component. However, it is not valid for the chromatic opposition components, as Figure 6.19b shows.

6.6.2. *The Effect on Chrominance*

At Figure 6.19b, we see that the amplitudes of the chromatic components decrease with increasing light intensity. This phenomenon corresponds to a desaturation of the perceived color. Another phenomenon is known to occur

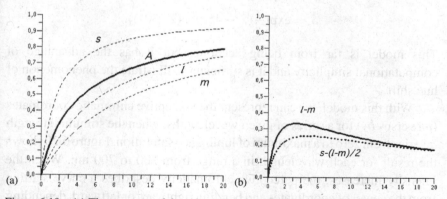

Figure 6.19. (a) The response of the achromatic component A with light intensity follows the same shape as the responses of photoreceptors, likely a Naka–Rushton law (see text). (b) The chromatic opposition components follow a quite different law.

when light intensity increases: the *Bezold-Brücke* effect (von Bezold, 1874; review in Imho *et al.*, 2004). It is a perceptual hue shift when the intensity of a stimulus is increased. A number of studies report a predictable hue shift with changes in intensity (Purdy, 1931; Boynton and Gordon, 1965; Cohen, 1975; Ejima and Takahashi, 1984). When intensity increases, long and middle wavelengths appear yellower and shorter wavelengths appear bluer.

The Bezold-Brücke effect has first been explained in terms of differential bleaching of photopigment, later (Vos, 1986) proposed a model of receptor adaptation.

In order to model this effect, let us come back to the photoreceptors adaptation law (equation 6.24) and consider the Red-Green and Blue-Yellow color opponent components, as seen in Chapter 5, with the cone proportions that have already been used:

$$lm = 5/16(l - m),$$
$$by = 15/16\,s - 10/16l - 5/16\,m. \tag{6.26}$$

We then include a simplified model of cone spectral sensitivity, using the following Gaussian functions:

$$L = \exp(-(\lambda - 580)^2/(2 \cdot 50^2)),$$
$$M = \exp(-(\lambda - 540)^2/(2 \cdot 50^2)), \tag{6.27}$$
$$S = \exp(-(\lambda - 450)^2/(2 \cdot 30^2)).$$

This model is far from being accurate, but it has the advantage of computational simplicity and it is sufficient to illustrate the phenomenon of hue shift.

With this model, we can represent the perceptive chromatic coordinates (*lm* versus *by*) for a series of given wavelengths, when the stimulus strength *a* varies from zero to a maximum of luminance saturation. Figure 6.20 shows the result for each wavelength in a range from 300 to 700 nm. When the intensity increases, the perceptive chromaticity describes a curve starting from the center of coordinates and bending rightward or leftward, depending on the region of the *lm*/*by* plan. The bending occurs in three regions: towards the yellow direction (560 nm), toward the blue direction, away from the

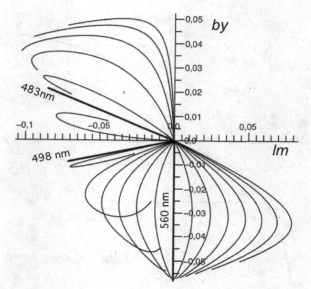

Figure 6.20. Model of the Bezold-Brücke effect. When the strength of light increases (0 at center of the graph), the perceived hue changes with high intensities and the perceived color may be desaturated.

green direction (498 nm). For these three particular values of wavelength (483 nm, 498 nm and 560 nm), no bending occurs.

This fact has been experimentally observed by Miyoshi *et al.* (1980): *"No hue shift was observed at around 570 nm over a whole range of the retinal illuminance at all durations employed and at around 480 nm at 300 ms duration"*. This observation reports only two wavelengths with no hue shift (our first and third cases). This is not surprising, because, in such a model, the number of solutions for no hue shift may vary, depending on the various cone proportions. As these proportions strongly vary from individual to individual, we are likely to observe different numbers and values of hue stability in each case.

6.6.3. *The Complementary Color After-Effect*

When we look at a colored surface during one or two minutes, the perceived color varies with the time of observation. After some time, the perceived color becomes more and more desaturated, up to a certain limit that depends on the intensity of the color. If we suddenly look at a white surface, it appears

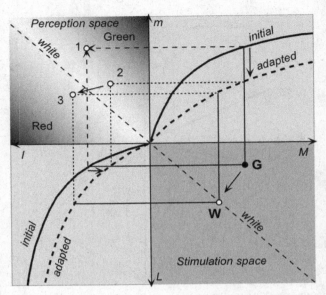

Figure 6.21. Complementary color after-effect. L, M: stimulation space. l, m: perception space after the non-linear transformations $l(L)$ and $m(M)$. A green stimulus is initially perceived as green (1). After cone adaptation, it becomes greenish (2). If the stimulus suddenly becomes white, the perception is pink (3).

to us as the complementary color! This is a phenomenon typically known to happen due to the photoreceptor adaptation process.

A simplified explanation is given by Figure 6.21. For simplicity's sake, let us imagine that we have only two receptor types: L and M. We define this stimulation space as the locus of points of coordinates (L,M). If $M >$ L, the stimulus color is Green. If $M = L$, the stimulus color is white. Each photoreceptor has a non-linear transduction function $l(L)$ and $m(M)$ respectively, which produce the perception space (l, m).

The non-linear transduction functions will adapt according to time and stimulus intensity (through their parameter L_0 and M_0), changing from an initial form to an adapted form after some time of stimulation. Suppose a Green stimulus (point G, Figure 6.21) applied at time $t = 0$: the perceived color is found from the initial transduction curves (plain curves and point 1) and looks green. After some time, both curves evolve (dashed curves), at a different rate, stronger for $m(M)$ than for $l(L)$ because $M > L$. The perceived color is then given by the adapted curves (point 2) and looks more or less greenish. If at this point the stimulus suddenly becomes white ($M = L$), the immediate perceived color is given by the adapted curves (point 3). It then looks reddish, the complementary color of greenish!

6.6.4. *Color Constancy*

If we look at the same scene but in different conditions, for example during sunset, during midday in full sun, or else, while wearing colored sunglasses, we can determine very accurately what the real colors of objects are. This phenomenon is called "(perceived) color constancy". Our eye (or brain) manages to recover the real color of objects, whatever the color of the illuminating light (Smithson and Zaidi, 2004).

A number of studies have attempted to develop predictive models for color perception (Hurlbert, 1998). Among them, a first approach aims at establishing models of possible neural circuits. For example, some models imply a gain control mechanism (von Kries, 1902; Burnham *et al.*, 1957; Webster and Mollon, 1995; Meylan *et al.*, 2007). Another approach, rather computational, deals with the problem as that of a black-box system that needs to be characterized (Brainard and Wandell, 1992; D'Zmura, and Iverson, 1993; Adelson and Pentland, 1996; Maloney and Yang, 2001; Mamassian *et al.*, 2002; Brainard *et al.*, 2003; Brainard *et al.*, 2006). Both approaches are not mutually exclusive and share some common principles as for example the famous RETINEX theory (Land and McCann, 1971).

Let us consider again the Naka–Rushton compression law for each photoreceptor i, as a function of the spatial variable s:

$$x_i(s) = \frac{X_i(s)}{X_i(s) + X_{i0}(s)},$$

the input signal of each photoreceptor being an integration over wavelength of the product of the local illumination $E(s, \lambda)$ by the local surface reflectance $\rho(s, \lambda)$, weighted by the spectral sensitivity $\varphi_i(\lambda)$:

$$X_i(s) = \int_\lambda \rho(s, \lambda) E(s, \lambda) \varphi_i(\lambda) d\lambda, \tag{6.28}$$

and as already seen, the term $X_{i0}(s)$ being the local spatial mean of $X_i(s)$:

$$X_{i0}(s) = \langle X_i(s) \rangle_s.$$

Let us decompose for each receptor the reflectance and the illumination into their mean value plus a zero-mean fluctuation:

$$\rho(s, \lambda) = \bar{\rho}_i(s, \lambda) + \tilde{\rho}_i(s, \lambda) \quad \text{and} \quad E(s, \lambda) = \bar{E}_i(s, \lambda) + \tilde{E}_i(s, \lambda),$$

with

$$\bar{\rho}_i(s) = \frac{\int_\lambda \rho(s, \lambda)\, \varphi_i(\lambda) d\lambda}{\int_\lambda \varphi_i(\lambda)\, d\lambda} \quad \text{and} \quad \bar{E}_i(s) = \frac{\int_\lambda E(s, \lambda)\varphi_i(\lambda) d\lambda}{\int_\lambda \varphi_i(\lambda) d\lambda}.$$

The product under the integral of Equation (6.28) can be expanded as:

$$\rho(s, \lambda)E(s, \lambda) = \bar{\rho}_i(s)\bar{E}_i(s) + \tilde{\rho}_i(s, \lambda)\bar{E}_i(s)$$
$$+ \tilde{E}_i(s, \lambda)\bar{\rho}_i(s) + \tilde{\rho}_i(s, \lambda)\, \tilde{E}_i(s, \lambda)$$

Knowing that the mean of the second and third terms is zero, the integral of Equation (6.28) gives:

$$X_i(s) = \bar{\rho}_i(s)\bar{E}_i(s) + \int_\lambda \tilde{\rho}_i(s, \lambda)\tilde{E}_i(s, \lambda)\varphi_i(\lambda)d\lambda,$$

that is, considering that the product $\tilde{\rho}_i(s, \lambda)\tilde{E}_i(s, \lambda)$ of two independent zero-mean functions is also zero-mean and likely of small value, we have the following approximation:

$$X_i(s) \approx \bar{\rho}_i(s)\bar{E}_i(s). \tag{6.29}$$

Then, the output signal of a photoreceptor takes the following form:

$$x_i(s) \approx \frac{\bar{\rho}_i(s)\bar{E}_i(s)}{\bar{\rho}_i(s)\, \bar{E}_i(s) + \langle \bar{\rho}_i(s)\bar{E}_i(s)\rangle_s}. \tag{6.30}$$

Considering that in usual conditions, the illumination $\bar{E}_i(s)$ is locally constant or at least, slowly variable, we can obtain the output signal as:

$$x_i(s) \approx \frac{\bar{\rho}_i(s)}{\bar{\rho}_i(s) + \langle \bar{\rho}_i(s)\rangle_s}. \tag{6.31}$$

This result shows that the photoreceptor's response depends almost entirely on the local reflectance, whatever the illuminant function is, provided it is slowly variable. In other terms, the photoreceptor's non-linear adaptation process exhibits the color constancy property. This fact has already been suggested by neuro-biologists considering the function of the horizontal

Figure 6.22. Example of the color constancy effect. Top left: the original image. Top right: the same image after the photoreceptors' adaptive compression. Bottom left: the original image seen through green glasses. Bottom right: the image after the photoreceptors' adaptive compression. The top and bottom right images represent the "perceived" images. They are rather similar in color. Inserts show the histograms of intensities. Hence the term of "color constancy" with respect to the color of the illuminant light. See colored figure in Color Plate ii.

cell to a cone feedback system (Kammermans *et al.*, 1998), providing some biological plausibility to our hypotheses.

From the standpoint of image processing, this principle provides an interesting algorithm in order to restore the "true" colors of a scene, independently of the illumination conditions. Figure 6.22 shows this type of example.

The left column of Figure 6.22 represents the same image seen normally (top) or through green glasses (bottom). The right column represents these images after the photoreceptors' adaptation process: first, they are much less different than the two images of the first column; second, their difference gives an idea of the amount of approximation in the Equations (6.29) and (6.30). An interesting aspect of this adaptive compression process is the fact that it widens the histograms of images (inserts in Figure 6.22), thus allowing a better extraction of the image information content (Atick, 1992; Chapeau-Blondeau, 1994; Nadal and Parga, 1994).

6.6.5. *MacAdam Ellipses*

In the 1940's, MacAdam undertook research on "Just Noticeable Differences" (JND) in color perception (MacAdam, 1942). He presented a colored surface to an observer and varied the color in different directions in the chromaticity diagram, and asked the observer to tell him when a difference in hue or saturation was perceived. The regions of JND drew ellipses of various elongations and orientations in the chromaticity diagram. The phenomenon was quite intriguing because the results varied from observer to observer, and with time, for a same observer. Many other research has been done later on the subject (e.g. Wyszecki and Fielder, 1971; Nagy and Eskew, 1987).

More recently, a generic model of color discrimination was presented in (Alleysson and Herault, 2001), involving adaptive nonlinearities at photoreceptor level and in color-opponent pathways. This model reproduces the various aspects of the data observed from six subjects, as reported by MacAdam, Wyszecki and Fielder. It is based on two main hypotheses:

- all the observers have the same kind of nonlinear adaptive functions,
- each observer has his or her own coding of color oppositions.

For each observer, the mean model parameters are adjusted to fit the experimental data in the particular available experimental conditions. In the model, one set of parameters depends on the adaptation state to light. The other set depends on the observer's specific color-coding scheme (e.g. relative proportions of photoreceptor types). With these parameters, the model faithfully reproduces the ellipses of MacAdam in almost all regions

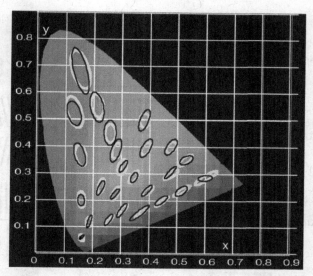

Figure 6.23. MacAdam ellipses. In black: experimental data. The ellipses represent the locus of just noticeable differences in color perception in the chromaticity diagram (ten times the real size). In white: the ellipses provided by the retinal model of non-linear adaptation.

of the chromaticity diagram (Figure 6.23). The result is that the observed variability in color discrimination ellipses stems only from:

- differences in the adaptation states, due to experimental conditions,
- inter-observer color coding differences.

6.7. Adaptation in the Visual Cortex

6.7.1. *Visual After-Effects*

Visual after-effects make up an important part of the psychophysics field of research. The study of these after-effects is expected to bring significant progress in the understanding of the neurophysiological basis of visual perception, as well bringing insight about the functional role they play in signal and information processing. Some examples are given in Figure 6.24. They concern adaptation phenomena to color (as already seen), orientation, spatial frequency or visual motion. For orientation and spatial frequency, the observer is asked to look at the horizontal bar during 30–60 seconds (induction phase) and then is presented with the pattern of the central

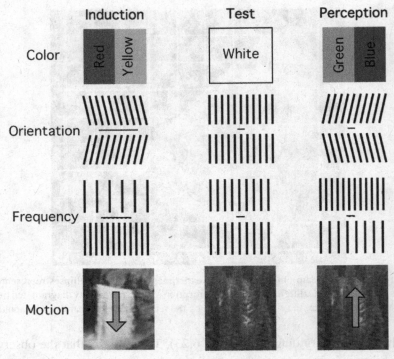

Figure 6.24. Visual after-effects. The general principle is the following. During a first phase (induction, left column), the subject looks at an image during 30–60 seconds. During a second phase (test, central column) a new image is presented and the subject is asked to say what pattern is perceived (right column).

column and asked to look at the central bar (test phase). The perception of the observer is given in the right column of Figure 6.24. In case there is motion, it is the well-known "water-fall effect". These effects have been widely studied in the past decades (e.g. Barlow, 1990; review in Kohn, 2007) and continue to be actively investigated.

In general, these after-effects share common properties: they are mono-dimensional (only one aspect of the signal is concerned) and they last for a short duration of a few seconds to one minute. They are different from other consecutive effects such as the MacCollough effect (see next section), which are multi-dimensional and may last for hours or days.

There have been many attempts to model the after-effects, from functional models to more biologically plausible ones based on neural

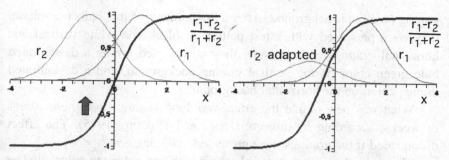

Figure 6.25. Visual after-effects. The combined response of two receptors may appear shifted, according to the adaptation state of one receptor.

adaptation processes (e.g. Drew and Abbott, 2006). Here is a simple model which is able to account for the basics of after-effects.

Suppose that a sensory dimension of value x is coded by several "receptors" scattered along the axis x as for example in Figure 6.25; two receptors are placed at neighboring positions on the x axis and exhibit bell-shaped responses r_1 and r_2 (thin curves on the figure). Now suppose a post-receptoral process, which provides a local measurement of x by:

$$r(x) = \frac{r_1(x) - r_2(x)}{r_1(x) + r_2(x)},$$

shown as a sigmoidal bold line on the figure. When the system is at rest (Figure 6.25, left) the response $r(x)$ is symmetric with respect to the receptors responses.

When a stimulus of some value x_0 is applied (grey arrow), after some duration, the receptor with the strongest response r_2 adapts more (Figure 6.25, right), and the post-receptoral process reveals a shifted response. Now, if a value of $x = 0$ is suddenly applied, the response is not zero any more, as in the situation at rest, but around 0.5, leading to a "perception" of the opposed sign to the induction signal.

6.7.2. *The McCollough Effect*

6.7.2.1. *Phenomenology*

In 1965, Celeste McCollough reported a particular color after-effect linked to orientation: subjects were presented alternatively every few seconds with a grating of vertical black stripes on an orange background, and a horizontal grating on a blue background. After a few minutes of this induction phase, they were presented with a test pattern of black and white vertical and horizontal gratings, side by side: they all reported seeing a de-saturated blue green color on the vertical grating background, and a de-saturated orange color, on the horizontal background.

What was new is that the effect was long lasting: up to hours, days or weeks, according to subjects (Jones and Holding, 1975). The effect disappeared if the gratings were presented at 45 degrees.

The same phenomenon was observed with red and green colors, and/or for patterns oriented at ±45°, or with other features (review in Allan and Siegel, 1997).

Furthermore, there exists a *reciprocity* of the McCollough effect between dimensions: for subjects who have looked at red stripes tilted clockwise off vertical and green stripes, tilted counterclockwise, the vertical test stripes appeared to be tilted counterclockwise when red but clockwise, when green (Held and Shattuck, 1971).

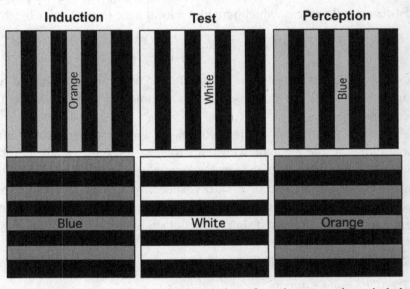

Figure 6.26. McCollough effect. Induction: during a few minutes, we alternatively look at Orange Vertical stripes and Blue Horizontal ones. Test: then we look at vertical and horizontal black and white stripes. Perception: we perceive colored stripes, vertical ones are blue and horizontal ones are orange, the exact opposite of the colors seen during the induction period. The effect can last for hours, days or weeks.

Despite a number of investigations and theoretical works, up to date the question of the locus of the mechanisms mediating the McCollough effect in the visual system is not completely clarified.

6.7.2.2. *Model of the McCollough effect*

Recently, Ans *et al.* (2001) proposed a model, which is able to account for all the above-mentioned properties. The model is based on the principle of blind separation of sources (BSS) or independent component analysis (ICA), a self-learning neural network developed in the 1980's (Hérault and Ans, 1984; Hérault and Jutten, 1986; Jutten and Hérault, 1991).

The neural network (Figure 6.27) is composed of two kinds of units. Two of them receive the Red-Green and Green-Red color signals as main inputs. The rest of them receive the orientation signals as main inputs. The outputs of the network are fed back in the following manner. Color outputs are fed back to the orientation units and orientation outputs are fed back

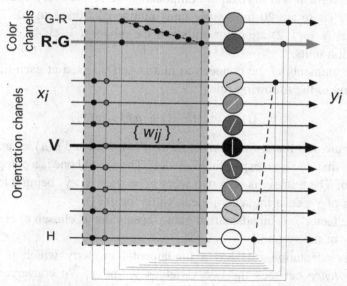

Figure 6.27. Neural network for the McCollough effect. The neurones inputs are the two color channels Red-Green and Green-Red, and the signals of orientation detectors. All the outputs of orientation neurones are fed back to the color channels inputs and the output of the two color neurones are fed back to the orientation channels, with a connecting matrix of weight w_{ij}.

to the color units, by means of a connection matrix $W = \{w_{ij}\}$. The neuron inputs are x_i. Their outputs are y_i.

The outputs are a weighted sum of inputs, combined with a non-linear activation function of the form:

$$y_i = N\left(x_i - \sum_{j \neq i} w_{ij} y_j\right). \tag{6.32}$$

The color of an external stimulus is simply characterized by a given amplitude (ranging from 0 to 1) of only one of the color components. In simulations, a red stimulus will be coded by ($R = 1$, $G = 0$) and a green one by ($R = 0$, $G = 1$).

The orientation inputs are assumed to originate from filters tuned to specific orientations. The 18 preferred orientations range from $-80°$ to $90°$ by $10°$, the filters responses are Gaussian (maximum 1, half-height bandwidth $25°$).

The vertical and horizontal components are respectively coded by $0°$ and $90°$. There are 20 processing units numbered from 1 to 20. The first two units ($i = 1, 2$) are color units and the others ($i = 3, 4, \ldots, 20$) are orientation units.

The elements of the connection matrix self-adapted at each time step according to the following rule:

$$w_{ij}(t) = w_{ij}(t - 1) + \mu f(y_i) g(\tilde{y}_j) \tag{6.33}$$

f and g are non-linear functions of form: $f(u) = u^3$ and $g(u) = \arctan(u)$.

The first one is an expanding function. The second one is a compressive function. The term $\tilde{y}_j$ is obtained from $\tilde{y}_j = y_j - \hat{y}_j$, $\hat{y}_j$ being a low-pass filtering of y_j, ensuring a zero-mean value for $g(\tilde{y}_j)$.

The factor μ is an adaptation gain whose value is chosen to ensure the stability of the system.

This adaptation rule has a fundamental property which is to *seek independence* between the two variables y_i and y_j. At convergence, the mean values of the connection matrix coefficients are zero:

$$\langle w_{ij}(t) - w_{ij}(t - 1)\rangle = 0,$$

that is: $\langle f(y_i)g(\tilde{y}_j)\rangle = 0$.

With the non-linear functions, the product $f(y_i)g(\tilde{y}_j)$ generates a series of odd powers of the form $y_i^{2p+1}\tilde{y}_j^{2k+1}$ and, the convergence will be obtained when all the joint moments of order $(2p+1)(2k+1)$ of the two variables y_i and y_j cancel. This is an approximative condition of independence between the variables. More in-depth theoretical approaches to independence can be found in (Pham *et al.*, 1992; Comon, 1994; Bell and Sejnowski, 1995; Charkani and Hérault, 1995; Comon and Jutten, 2007; Jutten and Comon, 2007).

6.7.2.3. *Simulations*

Induction. Two input patterns are alternatively presented to the network: the red-vertical input (R = 1, G = 0, $\theta = 0°$) and the green-horizontal input (R = 0, G = 1, $\theta = 90°$). During this phase, the weights (initially set to zero) are updated according to the adaptive rule 6.33. After an induction time $t = 5000$, the color outputs of the network are analyzed.

Test. Two test patterns are presented: either the achromatic-vertical (R = 0, G = 0, $\theta = 0°$) or the achromatic-horizontal (R = 0, G = 0, $\theta = 90°$). The results are as follows:

- For the achromatic-vertical test pattern the outputs are ($y_r = 0$, $y_g = 0.189$), and symmetrically for the achromatic-horizontal test pattern the color outputs are ($y_r = 0.189$, $y_g = 0$), that is, the opposite of the induction condition, as observed in the McCollough effect with human subjects.
- Considering that before the induction phase the maximum activity reached by a color unit is 0.632, the color after-effect that emerges from the network in test is roughly 30% of the maximum color value, that is, a de-saturation. This also is observed in the McCollough effect with human subjects.
- If the achromatic test grid is tilted from the inducting orientation, the network color response vanishes when the orientation test reaches 45°. Again, this is the same as the observations obtained with human subjects (McCollough, 1965).

Duration of the effect. Once adapted, the network is faced with input patterns reflecting a continually changing "ecological environment": To simulate a

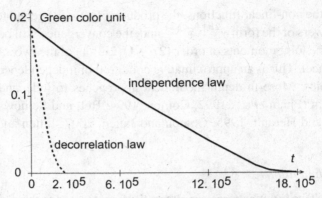

Figure 6.28. The decrease of the simulated McCollough effect: observe the very long persistency obtained with the independence law, versus the decorrelation law.

desadaptation phase, the network receives input patterns with colors and orientations changing randomly with time. From time to time, the test pattern is applied and the response of color units are recorded in order to assess the "memory" of the network.

Figure 6.28 represents the variation of the green color unit activity, in the course of the desadaptation phase, when the network is tested with the achromatic-vertical input pattern. We observe that, from the adaptation value ($y_g = 0.189$), the magnitude of the after effect decreases very slowly with time, and vanishes at $t = 18.10^5$, which corresponds to 360 times the induction time ($t = 5000$). Obviously, the same effect holds for the decrease of the red color after-effect when the network is tested with the achromatic-horizontal input pattern. Again, this behavior resembles the forgetting effect with human subjects.

It is interesting to see what happens if the adaptation law is a decorrelation rule (Equation 6.34), as the one used in (Barlow and Foldiac, 1989):

$$w_{ij}(t) = w_{ij}(t-1) + \mu y_i \tilde{y}_j. \tag{6.34}$$

In this case, the network desadaptation process is much faster (about 10 times) with respect to the independence rule. Though decorrelation has often been presented as a fundamental brain property, this independence rule seems to be more adequate and may be a potential model for brain functions and for the memorization processes. In fact, it is known in signal processing that statistical independence is far better than simple decorrelation.

A reverse after-effect: from color to orientation has been reported by Held and Shattuck (1971): during the inducting phase, subjects were presented red stripes tilted clockwise off vertical and green stripes tilted counterclockwise. In the test phase, the vertical stripes were perceived tilted counterclockwise when red but clockwise when green. Here again, the network is able to reproduce this effect (Ans *et al.*, 2001).

Chapter 7

Cortical Processing of Images

Since the pioneer of Hubel and Wiesel (1963, 1968), the biological principles of low-level vision are rather well identified. In the primary visual area V1, the retinal signals are filtered by simple (in-phase and in- quadrature) and complex (energy dependent) cells. Each cell responds to one central spatial frequency and one orientation. They are organized into micro-columns of similar frequency and orientation. There are approximately 6–7 central frequencies (on a log-scale) and 15–16 orientations (between 0° and 180°). Micro-columns are organized in a pin-wheel fashion within a macro-column (see review in Spillmann and Wemer, 1990). To put it briefly, after retinal preprocessing, the task of V1 complex cells can be viewed as a log-polar sampling of the energy spectra of local patches by band-pass oriented filters over the whole visual field (Pollen and Ronner, 1983).

Our fantastic ability to rapidly recognize a scene (within a few tens of milliseconds) raises the question of how the visual system encodes images (Potter and Faulconer, 1975; Thorpe et al., 1996; Oliva and Torralba, 2001; Oliva, 2005). In order to understand the principles of such a performance, we will first need to analyze the characteristics of natural images issued from the visual scenes, seen from various view-points.

In this chapter, we will see that the statistics of natural images is linked to their representation in terms of spatial frequency, through the Fourier transform. For this reason, the coding of images in V1 seems to be well adapted to this representation. After examining this first approach, we will show some models of cortical filters and various applications to scene categorization, as well as attention and monocular perspective estimation.

7.1. Properties of Natural Images

7.1.1. *Statistics of Natural Scenes*

A given image will be considered as a particular realization of a random process, which means that it is impossible to say what the exact value is for a given pixel. In other words, an image cannot be described by a mathematical formula. The only way, which gives access to image properties, is statistical analysis.

Let us consider a sample of N images drawn at random from a database.

First order statistics will give some general information such as:

- the statistical mean: $\mu(x, y) = \frac{1}{N} \sum_n i_n(x, y)$, or written as the mathematical expectation: $\mu(x, y) = \mathbf{E}[i(x, y)]$.
- the variance: $\sigma^2(x, y) = \mathbf{E}[((i(x, y) - \mu(x, y))^2]$.

Second order statistics will give information about the spatial relations between two pixels located at (x_1, y_1) and (x_2, y_2).

- The statistical autocorrelation is:

$$R(x_1, y_1, x_2, y_2) = \mathbf{E}\left[i(x_1, y_1) \cdot i(x_2, y_2)\right]. \qquad (7.1)$$

In this case, the statistics can be computed over all possible images or within a specific image category (see examples, Figure 7.1).

A process is said to be *stationary* if its second order statistics depends only on the displacement $x = x_1 - x_2$, $y = y_1 - y_2$. In this case, the autocorrelation function is simpler:

$$R(x, y) = \mathbf{E}\left[i(x_1, y_1) \cdot i(x_1 - x, y_1 - y)\right]. \qquad (7.2)$$

The *theorem of Wiener-Kinchine* says that the Fourier transform of the autocorrelation function gives the spectral power density function of a stationary random process: $F\{R(x, y)\} = S(f_x, f_y)$. For natural images, its shape is known to be in $1/f^\alpha$, with $2 < \alpha < 3$ (Ruderman, 1994; Dong and Atick, 1995; Guérin-Dugué and Oliva, 2000; Simoncelli and Olshausen, 2001). To be more precise, the power coefficient depends on the orientation θ: $S \approx 1/f^{a(\theta)}$.

A random process is said to be *ergodic* if its statistical autocorrelation function $R(x, y)$ equals the spatial autocorrelation function $\gamma(x, y)$ of any

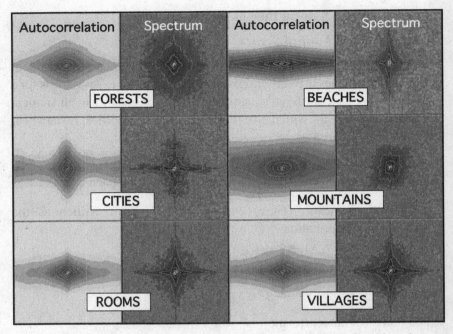

Figure 7.1. Various image categories, their statistical autocorrelation function and power spectrum.

particular realization:

$$R(x, y) = \gamma(x, y) \triangleq \iint_{x_1, y_1} i(x_1, y_1) \cdot i(x_1 - x, y_1 - y) \, dx_1 dy_1. \quad (7.3)$$

As the Fourier transform of the spatial autocorrelation function gives the squared amplitude spectrum, it becomes possible, under the ergodicity hypothesis, to compare the spectrum of a given image to the power spectrum of the database: $S(f_x, f_y) \Leftrightarrow A_n^2(f_x, f_y)$, or to the one of a given category: $S_c(f_x, f_y) \Leftrightarrow A_n^2(f_x, f_y)$, see Torralba and Oliva (2003).

7.1.2. *Fourier Transform and Frequency Spectra of Images*

7.1.2.1. *Image Frequency Spectra*

Let us now examine some properties of the Fourier transform and of the power spectrum of an image. For a given image $i(x, y)$, its Fourier transform is:

$$i(x, y) \rightarrow I(f_x, f_y) = \iint_{x,y} i(x, y) \, e^{-j2\pi(xf_x + yf_y)} dx dy, \qquad (7.4)$$

or in vector notation:

$$i(\mathbf{x}) \rightarrow I(\mathbf{f}) = \iint_{\mathbf{x}} i(\mathbf{x}) \, e^{-j2\pi \mathbf{x}^T \mathbf{f}} d\mathbf{x},$$

with $\mathbf{x} = (x, y)$ and $\mathbf{f} = (f_x, f_y)$.

The Fourier transform is a complex number written with a real and an imaginary part or with an amplitude and a phase:

$$I(\mathbf{f}) = \mathrm{Re}(\mathbf{f}) + j\mathrm{Im}(\mathbf{f}),$$
$$I(\mathbf{f}) = A(\mathbf{f}) \, e^{j\phi(\mathbf{f})},$$

its amplitude being $A(\mathbf{f}) = \sqrt{\mathrm{Re}^2(\mathbf{f}) + \mathrm{Im}^2(\mathbf{f})}$ and its phase being $\phi(\mathbf{f}) = \arctan\left(\mathrm{Im}(\mathbf{f})/\mathrm{Re}(\mathbf{f})\right)$.

7.1.2.2. *Manipulations on Image Spectra*

It is possible to build an image with a constant (flat) power spectrum by dividing its Fourier transform by its amplitude spectrum:

$$I_F(\mathbf{f}) = \frac{\mathrm{Re}(\mathbf{f})}{\sqrt{\mathrm{Re}^2(\mathbf{f}) + \mathrm{Im}^2(\mathbf{f})}} + j \frac{\mathrm{Im}(\mathbf{f})}{\sqrt{\mathrm{Re}^2(\mathbf{f}) + \mathrm{Im}^2(\mathbf{f})}} = e^{j\phi(\mathbf{f})}.$$

It is also possible to make a hybrid image with the amplitude spectrum of one image and the phase spectrum of another one:

Consider two images with spectra $I_1(\mathbf{f}) = \mathrm{Re}_1(\mathbf{f}) + j\mathrm{Im}_1(\mathbf{f})$ and $I_2(\mathbf{f}) = \mathrm{Re}_2(\mathbf{f}) + j\mathrm{Im}_2(\mathbf{f})$. The hybrid image is built from the amplitude spectrum of image #1 with the phase of image #2:

$$I_H(\mathbf{f}) = \frac{\mathrm{Re}_2(\mathbf{f}) + j\mathrm{Im}_2(\mathbf{f})}{\sqrt{\mathrm{Re}_2^2(\mathbf{f}) + \mathrm{Im}_2^2(\mathbf{f})}} \sqrt{\mathrm{Re}_1^2(\mathbf{f}) + \mathrm{Im}_1^2(\mathbf{f})}.$$

7.1.2.3. *Image Spectra Under Affine Transforms*

A scene can be seen from different points of view. The observer can be closer or farther, the observer can rotate his head. . . All these situations correspond to an affine coordinate transform. Furthermore, the center of sight can be at different locations in the scene, resulting in various translations of the retinal image. Let us see how the power spectrum of an image behaves under such conditions.

Affine coordinate transforms.
Let $\mathbf{A}$ be a matrix of coordinates transform: $\mathbf{x} \rightarrow \mathbf{A}\mathbf{x}$ and $i(\mathbf{x})$ an image. The Fourier transform of the transformed image $i'(\mathbf{x}) = i(\mathbf{A}\mathbf{x})$ is:

$$I'(\mathbf{f}) = \frac{1}{|Det(\mathbf{A})|} I(\mathbf{A}^{-T}\mathbf{f}). \tag{7.5}$$

- If the coordinates transform is an inversion of axes $(x, y) \rightarrow (-x, -y)$, the real part of the Fourier spectrum is unchanged and the imaginary part changes of sign. However, the amplitude and the power spectra do not change (The Power spectrum has a central symmetry with respect to $\mathbf{f} = 0$).
- If the coordinates transform is a zoom $(x, y) \rightarrow (ax, ay)$, the spectrum becomes $I'(\mathbf{f}) = 1/a^2 I(\mathbf{f}/a)$. The effect is an inverse zoom on the spectrum (up to a coefficient of $1/a^2$).
- If the coordinates transform is a rotation of angle θ, the frequency spectrum is rotated with the same angle.

Translations.
If an image is submitted to a translation $\Delta\mathbf{x}$, its Fourier transform is only multiplied by a complex number:

$$F\{i(\mathbf{x} + \Delta\mathbf{x})\} = F\{i(\mathbf{x})\} \cdot e^{j2\pi \mathbf{f}^T \Delta\mathbf{x}}.$$

This results in an important property: *the amplitude and power spectra are invariant with respect to spatial translations.*

As a scene category does not change under translation and affine transforms, the amplitude spectrum is a good candidate for image coding, as we will see later.

7.2. Cortical Filters Models and Spatial Frequency Analysis

7.2.1. *Simple and Complex Cells in V1 and Models*

Since Hubel and Wiesel (1974), it is well known that the primary visual cortex comprises orientation and frequency sensitive cells: the two types of simple cells (in-phase and in-quadrature) correspond mainly to the real and imaginary parts of the Fourier spectrum. Complex cells are related more to the amplitude spectrum. That is, they provide some measure of the local energy in the image. Complex cells comprise a large percentage of the units in V1, with estimates ranging from around 40–90% of the total (De Valois *et al.*, 1982; De Valois, 1991; Schiller *et al.*, 1976), which suggests an optimal coding of images, invariant with respect to translation.

Many filter shapes have been proposed since 1980 to perform an image transformation equivalent to that of simple cells (review in Wallis, 2001): the difference between two or more Gaussian functions – DoG – (Hawken and Parker, 1987), Laplacian of a Gaussian (Marr and Hildreth, 1980), Gabor (Marcelja, 1980; Daugman, 1980; Sakitt and Barlow, 1982; Jones and Palmer, 1987), Cauchy, the 'log Gabor' (Field, 1987; Morrone and Burr, 1988), and Log-Normal (Knutsson *et al.*, 1994; Massot and Hérault, 2008).

7.2.2. The Gabor Filter Model

Gabor's original work (Gabor, 1946) proposes a filter for one-dimensional signals that optimizes the tradeoff between time and frequency representations. Daugman (1980, 1992) extended this idea to two-dimensional images, for the trade-off between spatial location and spatial frequency: the two-dimensional Gabor filter represents an optimal combination between frequency and space information. This theoretical work has been very popular in neurophysiological studies (Kulikowski *et al.*, 1982; Jones and Palmer, 1987) as well as in psychophysical studies (Harvey and Doan, 1990).

7.2.2.1. *Bi-dimensional Gabor Filter and Local Frequency Analysis*

In one dimension, the frequency representation of a Gabor filter is made of a Gaussian envelope centered on a central frequency:

$$G(f) = e^{-\frac{(f-f_0)^2}{2\sigma^2}}. \tag{7.6}$$

The impulse response is complex. Its real part is a cosine modulated by a Gaussian envelope. Its imaginary part is a sine modulated by the same

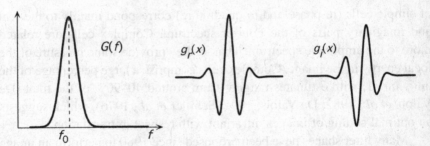

Figure 7.2. One-dimensional Gabor filter. Left: frequency representation. Right: real part (in-phase) and imaginary part (in-quadrature) impulse responses.

Gaussian envelope:

$$g_r(x) = \frac{\sigma}{\sqrt{2\pi}} \cos(2\pi f_0 x) \cdot e^{-\frac{\sigma^2 x^2}{2}}, \quad g_i(x) = \frac{\sigma}{\sqrt{2\pi}} \sin(2\pi f_0 x) \cdot e^{-\frac{\sigma^2 x^2}{2}}.$$

(7.7)

The real part is said to be "in-phase" and the imaginary part is said to be "in-quadrature", due to the $\pi/2$ difference of phase between cosine and sine functions (see Figure 7.2).

In the Gabor model of cortical filters, $g_r(x)$ and $g_i(x)$ correspond to the two main classes of simple cells. The model for complex cells is more commonly described by $|G(f)|^2$ in the frequency domain. In the spatial domain, it is a non-linear function, the sum of the squared outputs of both in-phase and in-quadrature filters.

In two dimensions, the formulation is slightly more complex. Let us consider a filter centered at frequencies $f_x = f_{x0}$ and $f_y = 0$. Its frequency response is:

$$G(f_x, f_y) = e^{-\frac{(f_x - f_{x0})^2}{2\sigma_x^2}} \cdot e^{-\frac{(f_y)^2}{2\sigma_y^2}}.$$

Its impulse response is also complex. Its real part is a cosine in x modulated by a Gaussian envelope in x and y. Its imaginary part is a sine in x modulated by the same Gaussian envelope:

$$g_r(x) = \frac{\sigma_x \sigma_y}{2\pi} \cos(2\pi f_0 x) \cdot e^{-\frac{\sigma_x^2 \sigma_y^2 x^2 y^2}{2}},$$

$$g_i(x) = \frac{\sigma_x \sigma_y}{2\pi} \sin(2\pi f_0 x) \cdot e^{-\frac{\sigma_x^2 \sigma_y^2 x^2 y^2}{2}}.$$

It is easy to generate the whole bank of filters: we just need to choose various center frequencies and apply various rotations in the (x, y) plan around the origin. This will result in the same rotations in the frequency plan.

7.2.2.2. Log-Polar Representation of Frequency

It is known since De Valois (1991) that the frequency bandwidth of cortical filters increases with their central frequency, whereas their orientation bandwidth remains almost constant (see Figure 7.4a). This suggests that if

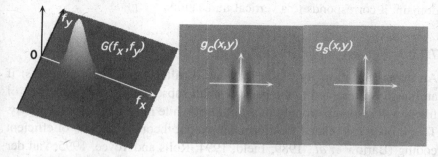

Figure 7.3. Two-dimensional Gabor filter. Left: frequency representation. Right: real part (in-phase) and imaginary part (in-quadrature) impulse responses.

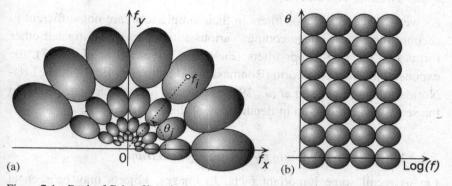

Figure 7.4. Bank of Gabor filters in the frequency domain. (a) in linear frequency axes, the filters look like a rose window, (b) in log-polar axes, they regularly sample the domain of log-frequencies and orientations.

the frequency bandwidth/central frequency ratio is constant, the progression of central frequencies is linear on a logarithmic scale. It is then more convenient to represent the frequency domain in a log-polar diagram. In this representation, the log-polar domain is regularly sampled by filters of constant size: the center frequencies f_i, as well as the orientations θ_j are equally spaced (Figure 7.4b).

This log-polar representation is particularly well suited to treat image zooms and rotations. If an image is zoomed ($\mathbf{x} \rightarrow a\mathbf{x}$), its frequency spectrum is also zoomed ($\mathbf{f} \rightarrow \mathbf{f}/a$) and the effect in log-polar domain is a horizontal translation of $-\log(a)$. Similarly, if the image is rotated, its frequency spectrum is rotated by the same angle, and in the log-polar domain, it corresponds to a vertical translation.

7.2.2.3. *Notion of Sparseness*

An interesting property of such a bank of filters is that of sparseness: if an image is presented to the bank, a small subset within the population of filters will respond strongly to a stimulus, while most will respond poorly. This property is related to information about theoretic notions of efficient coding (Barlow *et al.*, 1989; Field, 1994; Rolls and Tovee, 1995; van der Schaaf and van Hateren, 1996). See a review in (Simoncelli, 2003).

7.2.3. *The Log-Normal Model*

As we will see, the Gabor filters in their simple form are not sufficient to account for a good image coding. Various authors have investigated other forms like the Log-Gabor filters (Field, 1987; Fischer *et al.*, 2007), the exponential Chirp Transform (Bonmassar and Schwartz, 1997), or the Log-Normal filters (Knutsson *et al.*, 1994; Knutsson and Andersson, 2003). In the sequel, we present an in-depth study of this question.

7.2.3.1. *Optimally Sampling the Frequency Spectrum*

Let us recall some important facts. In images, objects may be seen at various positions, sizes or orientations. Their coding scheme should take into account these variations or free itself from them. Considering statistical aspects, the spectral energy density $i(x, y) \rightarrow S(f_x, f_y)^1$ is a good representation, being linked to the second order statistics. Furthermore, it can be independent of (moderate) variations in position. As the objects may vary independently in size and orientation, it is important to use log-polar coordinates:

$$S(f_x, f_y) \rightarrow S\left(\ln\left(\frac{f}{f_0}\right), \varphi\right), \quad \text{with } f = \sqrt{f_x^2 + f_y^2}$$
$$\text{and } \varphi = \arctan\left(\frac{f_y}{f_x}\right).$$

As in V1, the power spectrum is sampled by a limited number of points after a smoothing by weighting functions $|G(f/f_i, \varphi - \varphi_j)|^2$ issued from spatial filters: $g_{ij}(x, y) \rightarrow G(f/f_i, \varphi - \varphi_j)$, which gives the samples:

$$C_{ij} = \int_{\varphi=0}^{2\pi} \int_{f=0}^{\infty} S(f, \varphi) \, |G(f/f_i, \varphi - \varphi_j)|^2 f df d\varphi. \tag{7.8}$$

As sizes and orientations are independent, the smoothing functions should be with separable radial and angular variables:

$$C_{ij} = \int_{\varphi=0}^{2\pi} |G_a(\varphi - \varphi_j)|^2 \int_0^{\infty} S(f, \varphi) \, |G_r(f/f_i)|^2 \, f df d\varphi \tag{7.9}$$

with

$$|G_a(\varphi - \varphi_j)|^2 = A_1 \left(\frac{1 + \cos(\varphi - \varphi_j)}{2}\right)^n = A_1 \cos^{2n}\left(\frac{\varphi - \varphi_j}{2}\right), \tag{7.10}$$

and

$$G_r^2(f/f_i) = A_2 \frac{1}{f^2} \exp\left(-\frac{1}{2}\left(\frac{\ln(f/f_i)}{\sigma_r}\right)^2\right). \tag{7.11}$$

A_1 and A_2 are normalization factors. The form of the angular filter has the advantage of being 2π-periodic, just like the image frequency spectrum, and very close to a Gaussian $N(\varphi_j, \sigma_a)$, with $n = -1/[4\ln(\cos(\sigma_a/2))]$.

---

[1] In order to avoid symbol multiplicity, we will use $S(f)$ for the generic form, $S(f_x, f_y)$ for the Cartesian form, $S(f, \varphi)$ for the polar form, and $S(v, \varphi)$ for the Log-polar form.

The radial filter has a Log-Normal form. After changes of variables: $v = \ln(f/f_0)$ and $v_i = \ln(f_i/f_0)$, it gives the following form for the coefficient C_{ij}:

$$C_{ij} = \int_{\varphi=0}^{2\pi} \left| G_a(\varphi - \varphi_j) \right|^2 \int_{v=-\infty}^{\infty} S(f_0 e^v, \varphi) \, \frac{1}{\sigma_r \sqrt{2\pi}}$$

$$\times \exp\left(-\frac{1}{2} \left(\frac{v - v_i}{\sigma_r} \right)^2 \right) \, dv d\varphi \qquad (7.12)$$

Observations:

- Due to the (true) Log-Normal form of the radial component

$$G_r(f) = \sqrt{A_2} \frac{1}{f} \exp\left(-\frac{1}{4} \left(\frac{\ln(f/f_i)}{\sigma_r} \right)^2 \right),$$

the $1/f$ factor disappears with the change of variables in the integral of Equation (7.9), leading to a fundamental property with respect to image scaling, as will be seen later.
- In literature, we find similar expressions named Log-Gabor (Field, 1987) or even Log-Normal (Knutsson, 1994, 2003), but none have this important $1/f$ factor.

7.2.3.2. *Log-Normal Filters and Image Zoom and Rotation*

A zoom factor of α and a rotation of angle θ applied to an image $i(x, y)$ transforms its power spectrum as $S(f, \varphi) \rightarrow \frac{1}{\alpha^4} S(\frac{f}{\alpha}, \varphi - \theta)$. The expression of C_{ij} becomes:

$$C_{ij}(\alpha, \theta) = \frac{A}{\alpha^4} \cdot \int_{\varphi=0}^{2\pi} G_a^2 \left(\varphi - (\varphi_j - \theta) \right)$$

$$\times \int_{v=-\infty}^{\infty} S(f_0 e^v, \varphi) \exp\left(-\frac{1}{2} \left(\frac{v - (v_i - \ln(\alpha))}{\sigma_r} \right)^2 \right) \, dv d\varphi.$$

That is, zooms and rotations are merely transformed into pure translations on the filters. Furthermore, these filters being made steerable if their number

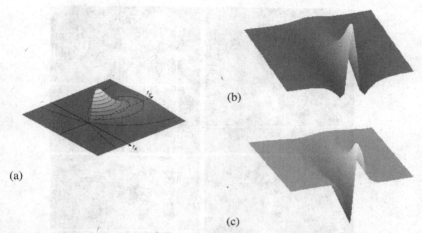

Figure 7.5. (a) spectral profile of a Log-normal filter in the frequency domain. (b) its in-phase impulse response, (c) its in-quadrature impulse response, in the spatial domain.

is odd, they allow either invariance with, or pursuit of zoom and rotation. This property would not be reached without the $1/f$ factor.

7.2.3.3. *Spectral Form and Coverage of Log-Normal Filters*

Because the frequency response of a Log-Normal filter is real, its 2-D impulse response has an in-phase part and an in-quadrature part (see Figures 7.5b and 7.5c).

Observations:

- In the frequency domain, the Log-Normal filter response is zero at $f = 0$. This is an important advantage with respect to the Gabor filter: even if its central frequency is low, it can accept any bandwidth, contrarily to the Gabor model.
- Even if its impulse response looks very similar to that of Gabor filters, its spectral behavior is quite different.
- The factor $1/f^2$ in $|G_r|^2$ is compatible with the $1/f^2$ shape of the power spectrum of natural images.
- The response profile of Log-Normal filters looks more like the biological data measured on V1 complex cells (Wallis, 2001), see Figure 7.6b.

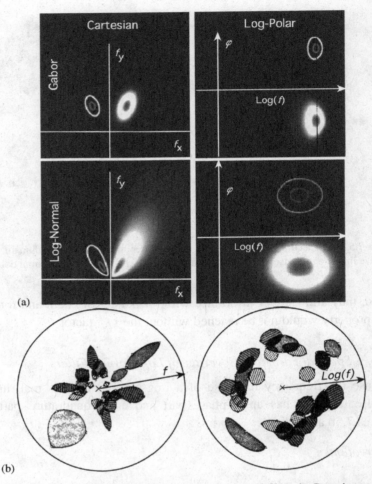

Figure 7.6. (a) comparison between Gabor and Log-Normal filters in Cartesian and log-polar frequency representation. (b) biological data on V1 complex cells in the frequency plane for linear (left) and logarithmic (right) radial frequencies (redrawn from De Valois *et al.*, 1982, with kind permission). Notice the similarity with Log-Normal filters.

- The spectral coverage of a bank of filters is more regular for Log-Normal filters than for Gabor filters (Massot and Hérault, 2008).

It should be noticed that the Log-Normal function has no simple analytic form (Leipnik, 1991), so the shape of the impulse response can be derived only by computer simulation.

7.3. Scene Categorization

7.3.1. *The Problem*

A scene is viewed within the spatial window of the field of vision. Recognizing the category of a scene in this spatial window implies some independency with respect to moderate changes in translation, zoom and rotation. For translation, a good candidate is the power spectrum of the image, and for zoom and rotation, the log-polar representation of frequency is more adequate, as we have previously seen. In the visual area V1, the large number of complex cells makes them good candidates for scene coding with respect to the above-mentioned image transformations. In other words, the set of complex cells in a macro-column samples the local power spectrum over frequencies and orientations.

7.3.1.1. *The Coding of Images*

The complex cells' outputs can be considered as the components of a vector with (number of frequencies × number of orientations) components. That is, in V1, each region of the retinal image is coded into a high-dimensional vector (see Figure 7.7) of dimension (number of frequencies × number of orientations × number of image regions).

If we consider the number of radial frequency bands to be around 6–7, the number of angular frequency bands between 0 and π around 7–8, and the number of macro-columns, around 1000 in each hemi-cortex, the dimension of the vector to code a scene would be around 50,000. This is quite a huge number. In a first step, in order to simplify in the purpose of categorization,

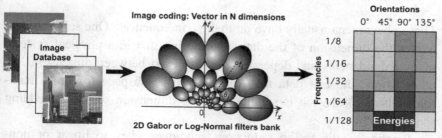

Figure 7.7. Coding images by a bank of 2-D filters into high-dimensional vectors, the components of which being the energy of the image in each frequency band. In this example, there are 5 frequency bands and 4 orientations.

we will consider a coding of the whole image by, for example, summing the local energies across all the different image regions. This will result in a vector of (number of frequencies × number of orientations) components.

In the sequel, we will consider 7 frequency bands and 7 orientations, leading to vectors in 49 dimensions. Hence, an image will be a point in a 49 dimensional space, and we hope that a category will be a cloud of points in this space. Furthermore, we hope that different categories correspond to different disjoint clouds.

7.3.1.2. *Representation of High-Dimensional Spaces*

Many problems occur when analyzing sets of numerical data high-dimensional spaces. From a theoretical point of view, several facts have to be considered:

- The "curse of dimensionality" (Bellman, 1957; Donoho *et al.*, 1998). This implies that the number of samples required to accurately approximate a data distribution grows exponentially with the dimensionality (Silverman, 1986).
- The empty space phenomenon (Scott and Thompson, 1983). The volume of a hypersphere inscribed in a hypercube rapidly tends to zero with respect to the volume of the hypercube, when the dimensionality grows.
- The problem of the metrics (Demartines and Hérault, 1997). It can be shown, given a random distribution of points in dimension N, that the mean of the Euclidean norm grows as $\sqrt{N}$, whereas the standard deviation tends to be constant. This implies that in high dimensions, all the inter-point distances tend to appear as equal!

These phenomena usually have dramatic consequences. One solution is to reduce the dimension of the data, considering that in a physical system, many redundancies and dependencies may exist between dimensions. In other words, the raw data may often lie in a subspace (or on a manifold) whose dimension is smaller than the dimension of the embedding space.

Practically, dimension reduction techniques obey to linear or non-linear models. Three well-known methods belong to the linear model class: Principal Component Analysis (Hotelling, 1933; Jolliffe, 1986), Projection

Pursuit (Friedman, 1987) and metric multidimensional scaling (Torgerson, 1952). Among the nonlinear models, we find the local PCA (Kambhatla and Leen, 1994), the Sammon's nonlinear mapping (Sammon, 1969), the Curvilinear Component Analysis (Demartines and Hérault, 1997; Hérault *et al.*, 1999; Guérin-Dugué *et al.*, 1999), the Isomap and the Curvilinear Distance Analysis (Lee *et al.*, 2004).

7.3.2. *Curvilinear Component Analysis*

In this section, we will see the Curvilinear Component Analysis algorithm, because it can treat highly non-linear and folded data structures.

7.3.2.1. *Principle*

Let us imagine a data structure like a blanket folded in 3-D space, with some noise, providing it with some thickness. In order to observe this blanket, we need only to unfold it on a 2-D plane. On this model, let us imagine that our data lie on a *p*-dimensional blanket, and are folded and embedded in a *n*-dimensional space. To observe these data, we must both unfold its structure and locally project them onto its mean manifold (see Figure 7.8). Then we must identify potential regions in this unfolded manifold. This process is known as "Non-Linear Mapping".

Let us consider an input space X of dimension n where the raw data are lying and an output (projection) space Y where the data are to be represented.

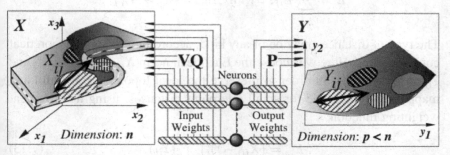

Figure 7.8. Principle of the CCA algorithm. The input weights first proceed to a vector quantization (VQ) of the input data space (X) in n dimensions. Then, the output weights map the local topology of the input average manifold by projecting it (P) into an output representation space (Y) of dimension $p < n$. This way, tasks like classification and recognition are highly facilitated in an unfolded and lower-dimensional output space.

An artificial neural network has N neurons. Each neuron is provided with an input weight vector x_i of dimension n pointing in the input space X and an output weight vector y_i of dimension $p < n$ pointing in the output space Y.

If the number of data points N_d is higher than the number of neurons N, a first step will consist in a vector quantization (VQ) of the data by the input weights. If the number of data points is not too high, each neuron vector is assigned to a data point: $N = N_d$.

Now, the idea is to project (P) the n-dimensional input vector in the output space, while preserving only the local topology of the input space. This way, a global unfolding of the data manifold is expected.

7.3.2.2. *Global Unfolding and Local Projection*

First, the output vectors are initialized at random in the output space. Then we select at random two data points i and j of the input space, estimate their distance X_{ij} and map this distance onto the two corresponding points in the output space. If the distance Y_{ij} is smaller than X_{ij}, the two points i and j in the output space are moved away. If Y_{ij} is larger than X_{ij}, the two points i and j in the output space are moved close together. This procedure is iterated across all pairs of data points.

The *unfolding part* of the algorithm follows the principle of a stochastic gradient descent. A cost function is chosen as follows:

$$E = \sum_{i,j} E_{ij}, \quad \text{with} \quad E_{ij} = \left(X_{ij} - Y_{ij} \right)^2.$$

The type of distances may be of any kind, according to various theoretical considerations. Here we choose the Euclidean norm $X_{ij} = \| x_i - x_j \|_2$ and $Y_{ij} = \| y_i - y_j \|_2$. In order to preserve the local topology, this unfolding mapping is only made for small interpoint distances, using a convenient weighting function:

$$E_{ij}^u = \left(X_{ij} - Y_{ij} \right)^2 F_\lambda(Y_{ij}), \tag{7.13}$$

with $F_\lambda(Y_{ij}) = 1$ for $Y_{ij} < \lambda$ and $F_\lambda(Y_{ij}) = 0$ for $Y_{ij} > \lambda$.

It has been proved that this weighting function, applied on the output distances, ensures a much better global unfolding than other mapping

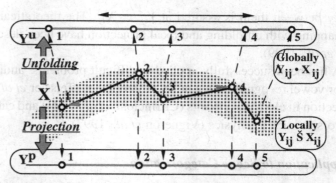

Figure 7.9. Illustration of the problem of data representation, in two cases: either only an unfolding is desired, or only a local projection is desired.

techniques, which apply it to the input distances (Demartines and Hérault, 1997).

The choice of λ strongly depends on the data structure (e.g. curvature of the average manifold, spreading of the data around this manifold). As the data structure is unknown in most cases, some strategy should be found, in order to define the best value of λ (see Demartines and Hérault, 1997).

We should observe that, apart from the desired global unfolding, there is also some tendency to make a local projection. Look at the input distribution in Figure 7.9: because we perform the mapping of X_{14} simultaneously with the mapping of X_{12}, X_{23} and X_{34}, the resulting compromise will lead to $Y_{12} < X_{12}$, $Y_{23} < X_{23}$ and $Y_{34} < X_{34}$, which is an approximative local projection. This property will be used later.

The *local projection part* of the algorithm also follows a kind of stochastic gradient descent, with a cost function chosen as:

$$E_{ij}^p = \left(X_{ij}^2 - Y_{ij}^2\right)^2. \tag{7.14}$$

It is important to notice that this is quite equivalent to making a local Principal Component Analysis, in the local subspace of the data. Because of the projection, this should apply only when $Y_{ij} \leq X_{ij}$, a situation which is initiated by the above-mentioned tendency to make local projection. Conversely, when $Y_{ij} \geq X_{ij}$, we are in the condition of unfolding. Hence, the two situations (unfolding or projection) do not overlap, and the global cost function can merge formulae (7.13) and (7.14), provided that the continuity between them is assured at $Y_{ij} = X_{ij}$. The theoretical aspects of this mapping with unfolding and local projection have been described in (Hérault *et al.*, 1999).

CCA has been successfully applied to difficult problems: audio-visual fusion for vowel recognition in a noisy environment (Teissier *et al.*, 1998), fault detection in electrical circuits (Cirrincione *et al.*, 1994) and calibration of detectors in nuclear physics (Vigneron *et al.*, 1997).

7.3.3. *Application to Scene Categorization*

Though the CCA algorithm has no known biological foundation, it can be interestingly used for scene categorization from the spatial statistics of the energy distribution of an image in various frequency bands and orientations. In (Hérault, Oliva and Guérin-Dugué, 1997) an image is analyzed by a bank of spatial filters, according to 4 orientations and 5 frequency bands, ranging from very low spatial frequencies to medium ones. The global energies of the 20 filters' outputs constitute the 20-dimensional measure space, and each image is a 20-vector in this space.

By CCA, we have found that a 2-dimensional representation was possible and that, in this space, the organization of the data was surprisingly in accordance with some semantic meaning (see Figure 7.10). The database was a set of 72 images representing six categories: Forests, Mountains, Beaches, Cities, Villages and Rooms, the spectra of which were given in Figure 7.1.

The observation of this figure shows some interesting organization: the natural and artificial scenes are separated, and a progression between "Open" and "Closed" landscapes is visible.

It is interesting to compare this mapping with the one obtained from a psychophysics experiment (Le Borgne *et al.*, 2003). Because CCA works on input distances, it does not require input vectors. A distance matrix is sufficient. In this experiment, human observers had to judge the similarity of 105 selected images presented on a computer display. The perceived similarity of each image with every other image of the base has been measured, and a distance matrix has been made.

The application of CCA on these distance data reveals a semantic clustering of the scenes (Figure 7.11), recalling the one observed in Figure 7.10, with (of course) a better separation of categories.

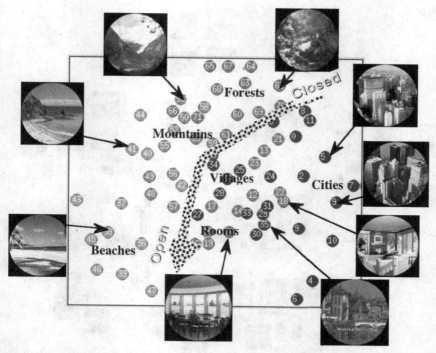

Figure 7.10. Two-dimensional representation of a 20-dimensional data space obtained by the energies, in 4 orientations and 5 spatial frequency bands, of a set of 72 images. We can clearly see that CCA reveals some semantic organization: Natural/Artificial scenes for each side of the grey line, and Open/Closed landscapes along this line.

7.4. Attention and Saliency

7.4.1. *Observing an Image*

Visual attentional processes are of two main types during the exploration of a scene: bottom-up when the ocular fixations are driven by the structure of the scene and top-down when the ocular fixations are driven by high-level processes linked to the semantic (or affective) content of the scene. The bottom-up process generally occurs for the first fixations during a naive visual exploration. Figure 7.12 provides an example of the density of fixation among a group of observers, without any given task. Only the first few fixations were registered for each observer (Chauvin *et al.*, 2002).

We remark that the subjects mainly look at two regions in the image: one with a boat and one with two persons. As the measurement concerns the first

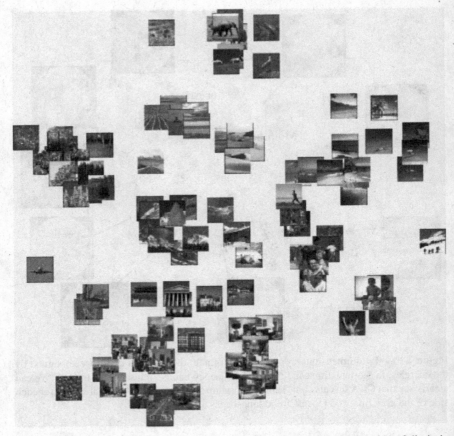

Figure 7.11.   Two-dimensional mapping of a database from human estimation of dissimilarities between images. The semantic clustering is obvious.

fixations in naïve conditions, we can expect that the attentional process is driven by the local structure: the fixated points have local statistics different from that of their surrounding regions. Given this fact, several authors have proposed various models of this bottom-up attentional process (Itti *et al.*, 1998; Chauvin *et al.*, 2002; Marendaz *et al.*, 2005).

7.4.2. *Modeling Saliency*

The Chauvin's model has an interesting biological plausibility: it takes into account a model of the retinal pre-processing of images with the

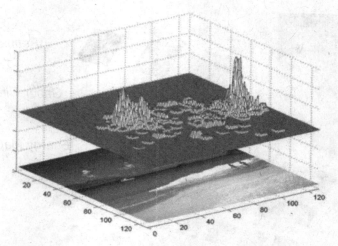

Figure 7.12. Map of the density of ocular fixations for an image observed by different human subjects (from Chauvin *et al.*, 2002, reprinted with kind permission).

Parvo- and Magnocellular pathways, and at the V1 level, the local and long-range cortical interactions (Polat and Sagi, 1999; Ben-Shahar *et al.*, 2003). After the interactions model, local texture contrasts maps, derived in several spatial frequency bands, help to build the global saliency map (Figure 7.13).

7.4.3. *Comparison with Human Behavior*

In order to verify the model's plausibility, psychophysical experiments have to be conducted. For this purpose, Marendaz *et al.* (2005) have manipulated the saliency and the semantic congruency of small regions within 72 natural scenes during a categorization task.

In the "saliency" experiment, the mean saliency (according to the model) of one region was reduced for half the scenes. In the "semantic" experiment, an incongruent object (object with a very low probability of occurrence in the scene context) was introduced. This object was substituted for another, but without any modification in the local power spectrum (hence, expected saliency). For both experiments, the luminance and contrast distribution of the modified regions were held constant. The third experiment consisted of a control condition (without any image

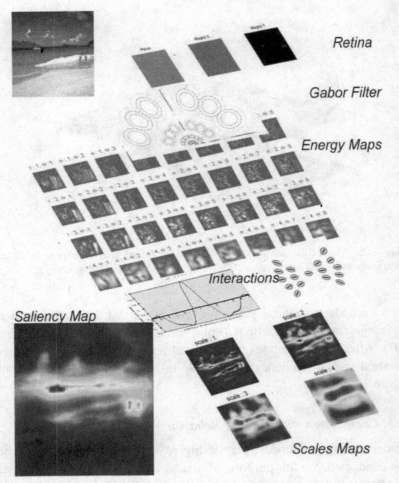

Figure 7.13.   Architecture of the model for the generation of saliency maps. (from Chauvin *et al.*, 2002, reprinted with kind permission.)

manipulation). Eye movements were recorded during the 3 seconds of the scene presentation and the number of fixations and the time spent inside the manipulated regions were registered.

In contrast with control condition, data analysis showed that, first, when the saliency of one region was reduced, the region was less attractive (the number of fixations and the time spent inside the region were lower than in the control condition). Secondly, when an incongruent object was inserted, in contrast to expectations according to some literature (Henderson and Hollingworth, 1998), the region was less attractive. This is true during the first moment of exploration, but this fact is reversed in the case of longer explorations.

7.5. Perspective Estimation

The projection of a viewed surface on the retina induces geometrical transformations to the texture covering this viewed surface. We show here how a 3-D information could be recovered by means of the cortical filters, using only the local energy spectrum of retinal images.

Since the seminal work of Gibson (1950), texture gradients have been considered as important monocular cues to estimate the orientation and shape of textured surfaces. Neurophysiological evidences (Tsutsui *et al.*, 2002; Liu *et al.*, 2004) as well as psychophysical ones (Knill, 1995; Li and Zaidi, 2004) support the idea of texture gradients processing, in the visual system.

Computationally, the problem deals with the estimation of two parameters: the inclination of the viewed surface in depth (slant) and the direction of the slant (tilt). In computer vision, since the early 90's, spatial frequency analyses have led to several efficient "shape from texture" algorithms. Some approaches (Super and Bovik, 1995a; Malik and Holtz, 1997; Sakai and Finkel, 1997; Guérin-Dugué and Elghadi, 1999; Ribeiro and Hancock, 2001; Loh and Kovesi, 2005) measure the texture deformation using the affine distortion of the pattern of spectral components (e.g. energy peaks, local moments, inertia). These methods are accurate only if they are in the presence of at least two orientations in regular textures. They are not appropriate in the case of irregular textures.

Alternative techniques do not make any assumption on the spectral components (Super and Bovik, 1995b; Garding and Lindeberg, 1996; Hwang *et al.*, 1998; Lelandais *et al.*, 2005). They choose the best local scale, explicitly assuming that there is only one at a given local spatial position. This is not always true in specific cases of multiple, occluded or transparent textures (Black and Rosenholtz, 1995). Being more or less efficient for irregular textures, they require heavy computation loads such as parabolic approximation or variance minimization.

Clerc and Mallat (2002), estimate the local migration of wavelet coefficients related to the local shape parameters, with interesting results on irregular textures. However, because it is an ergodicity hypothesis, the central frequency of each wavelet is relatively high and does not take into account the advantage of having the whole range of spatial frequencies.

None of these techniques can be compared to biological visual processing, except the technique of Sakai and Finkel (1997), who proposed a model with the ability to handle peak and mean variations of spatial frequency.

7.5.1. *The Geometry of a Conic Projection*

To compute the geometry, let us first place the viewed surface at the origin, rotate it with an angle σ (slant) around the x_w axis, and then rotate it with an angle τ (tilt) around the z_w axis (Figure 7.13). The world coordinates of the viewed surface become:

$$\begin{vmatrix} x_w \\ y_w \\ z_w \end{vmatrix} = \begin{bmatrix} \cos(\tau) & \sin(\tau) & 0 \\ -\sin(\tau) & \cos(\tau) & 0 \\ 0 & 0 & 1 \end{bmatrix} \cdot \begin{bmatrix} 1 & 0 & 0 \\ 0 & \cos(\sigma) & \sin(\sigma) \\ 0 & -\sin(\sigma) & \cos(\sigma) \end{bmatrix} \cdot \begin{vmatrix} x_s \\ y_s \\ z_s \end{vmatrix}.$$

$$(7.15)$$

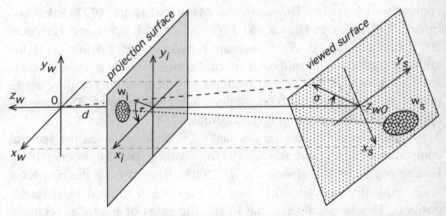

Figure 7.14. The origin of world coordinates $(x, y, z)_w$ is the center of projection. The viewed surface is at a position $-z_{w0}$ from origin, it is slanted (σ) and tilted (τ). Its coordinates $(x, y)_s$ correspond to $(x, y)_i$ on the projection surface. We will consider local information within two corresponding windows w_i and w_s (see text).

As the viewed surface is a plane, $z_s = 0$. Then, let us translate it to its place at z_{w0} and calculate the projection on the image plane:

$$x_i = \frac{d}{z_{w0} - z_w} x_w, \quad y_i = \frac{d}{z_{w0} - z_w} y_w,$$

which gives:

$$\begin{vmatrix} x_i \\ y_i \end{vmatrix} = \frac{d}{z_{w0} + \sin(\sigma)\, y_s} \begin{bmatrix} \cos(\tau) & \sin(\tau)\cos(\sigma) \\ -\sin(\tau) & \cos(\tau)\cos(\sigma) \end{bmatrix} \cdot \begin{vmatrix} x_s \\ y_s \end{vmatrix}. \qquad (7.16)$$

From this, we can calculate the inverse transform:

$$\begin{vmatrix} x_s \\ y_s \end{vmatrix} = \frac{z_{w0}}{d - \tan(\sigma)\,[\sin(\tau)x_i + \cos(\tau)y_i]}$$
$$\times \begin{bmatrix} \cos(\tau) & -\sin(\tau) \\ \sin(\tau)/\cos(\sigma) & \cos(\tau)/\cos(\sigma) \end{bmatrix} \cdot \begin{vmatrix} x_i \\ y_i \end{vmatrix}.$$

In vector-matrix notation:

$$\mathbf{x}_s = \frac{1}{a_i}\, \mathbf{A} \cdot \mathbf{x}_i, \qquad (7.17)$$

with:

$$\mathbf{x}_i = \begin{bmatrix} x_i & y_i \end{bmatrix}^{\mathrm{t}}, \quad \mathbf{x}_s = \begin{bmatrix} x_s & y_s \end{bmatrix}^{\mathrm{t}},$$
$$a_i = \{d - \tan(\sigma)\,[\sin(\tau)x_i + \cos(\tau)y_i]\}/z_{w0}.$$

7.5.2. *Local Spatial Frequency Spectra*

We will now compare the spatial frequency on the projection image, with the spatial frequency on the viewed surface. Let us consider a local window w_i centered at the position x_i in the image, corresponding to a local window w_s centered at the position x_s on the viewed surface.

The local frequency spectrum in the image is, in matrix-vector notation:

$$I_{Li}(\mathbf{f}_i, \mathbf{x}_i) = \int_{\mathbf{u}} i_i(\mathbf{x}_i + \mathbf{u})\, w_i(\mathbf{u})\, e^{-j2\pi \mathbf{u}^{\mathrm{t}} \mathbf{f}_i}\, d\mathbf{u}. \qquad (7.18)$$

Considering an ideal projection with no intensity attenuation, we can state:

$$i_i(\mathbf{x}_i + \mathbf{u})\, w_i(\mathbf{u}) = i_s(\mathbf{x}_s + \mathbf{v})\, w_s(\mathbf{v}) = \int_{\mathbf{f}_s} I_{Ls}(\mathbf{f}_s, \mathbf{x}_s)\, e^{+j2\pi \mathbf{v}^{\mathrm{t}} \mathbf{f}_s}\, d\mathbf{f}_s.$$

Then combining it with the preceding formula gives this:

$$I_{Li}(\mathbf{f}_i, \mathbf{x}_i) = \int_{\mathbf{u}} \int_{\mathbf{f}_s} I_{Ls}(\mathbf{f}_s, \mathbf{x}_s)\, e^{-j2\pi(\mathbf{u}^t\mathbf{f}_i - \mathbf{v}^t\mathbf{f}_s)}\, d\mathbf{f}_s d\mathbf{u}. \qquad (7.19)$$

One could think that the relation between $\mathbf{v}$ and $\mathbf{u}$ is the same one as between $\mathbf{x}_s$ and $\mathbf{x}_i$. In fact, because of the non-linearity, we should take the Taylor series expansion. By limiting it to the first order:

$$\mathbf{v} = \frac{1}{a_i}\mathbf{A}\left[I - \frac{\mathbf{x}_i \nabla_a^t}{a_i} \right] \mathbf{u}, \qquad (7.20)$$

that is:

$$
\begin{aligned}
I_{Li}(\mathbf{f}_i, \mathbf{x}_i) &= \int_{\mathbf{f}_s} I_{Ls}(\mathbf{f}_s, \mathbf{x}_s) \\
&\quad \times \int_{\mathbf{u}} \exp\left(-j2\pi\, \mathbf{u}^t \left(\mathbf{f}_i - \frac{1}{a_i}\left[I - \frac{\nabla_a \mathbf{x}_i^t}{a_i} \right] \mathbf{A}^t \mathbf{f}_s \right) \right) d\mathbf{u}\, d\mathbf{f}_s \\
&= \int_{\mathbf{f}_s} I_{Ls}(\mathbf{f}_s, \mathbf{x}_s)\, \delta\left(\mathbf{f}_i - \frac{1}{a_i}\left[I - \frac{\nabla_a \mathbf{x}_i^t}{a_i} \right] \mathbf{A}^t \mathbf{f}_s \right) d\mathbf{f}_s. \qquad (7.21)
\end{aligned}
$$

This means that the local frequency in the projection image is related to the local frequency on the viewed surface by:

$$\mathbf{f}_i = \frac{1}{a_i}\left[I - \frac{\nabla_a \mathbf{x}_i^t}{a_i} \right] \mathbf{A}^t \mathbf{f}_s, \qquad (7.22)$$

Assuming that the local frequency on the viewed surface is constant within a given area, the spatial variation of the local frequency in the image is only due to the perspective components σ and τ. After some calculus, we derive:

$$d\mathbf{f}_i = -\frac{\nabla_{a_i}^t d\mathbf{x}_i}{a_i}\, \mathbf{f}_i - \frac{\nabla_{a_i} d\mathbf{x}_i^t}{a_i}\, \mathbf{f}_i, \qquad (7.23)$$

Let us now consider the spatial frequency plane in polar coordinates $(\nu_i,\ \varphi_i)$ with $\mathbf{f}_i = \nu_i \left| \cos(\varphi_i)\ \ \sin(\varphi_i) \right|^t$. We have:

$$d\mathbf{f}_i = d\nu_i \left| \begin{matrix} \cos(\varphi_i) \\ \sin(\varphi_i) \end{matrix} \right| + \nu_i \left| \begin{matrix} -\sin(\varphi_i) \\ \cos(\varphi_i) \end{matrix} \right| d\varphi_i.$$

By multiplying $d\mathbf{f}_i$ by $\left| \cos(\varphi_i) \quad \sin(\varphi_i) \right|^t$, we obtain the radial component (modulus) of the local frequency:

$$
dv_i = -\frac{\nabla_{a_i}^t d\mathbf{x}_i}{a_i} v_i - \frac{1}{a_i} \left[\cos(\varphi) \quad \sin(\varphi) \right] \nabla_{a_i} d\mathbf{x}_i^t \begin{bmatrix} \cos(\varphi) \\ \sin(\varphi) \end{bmatrix} v_i,
$$

that is, dividing both members by v_i:

$$
d\log(v_i) = -\frac{\nabla_{a_i}^t d\mathbf{x}_i}{a_i} - \frac{1}{a_i} \nabla_{a_i}^t \begin{bmatrix} \cos^2(\varphi) & \sin(\varphi)\cos(\varphi) \\ \sin(\varphi)\cos(\varphi) & \sin^2(\varphi) \end{bmatrix} d\mathbf{x}_i^t.
$$
(7.24)

By taking the mean value over φ between 0 and 2π, we have[2]:

$$
\langle d\log(v_i) \rangle = \frac{1}{2\pi} \int_0^{2\pi} d\log(v_i) \, d\varphi = -\frac{3}{2} \frac{\nabla_{a_i}^t d\mathbf{x}_i}{a_i},
$$

that is:

$$
\langle d\log(v_i) \rangle = \frac{3}{2} \tan(\sigma) \frac{\sin(\tau)dx_i + \cos(\tau)dy_i}{d - \tan(\sigma)\left[\sin(\tau)x_i + \cos(\tau)y_i\right]}.
$$
(7.25)

It is interesting to observe that this expression is *independent* of the distance z_{w0} between the observer and the fixation point.

The tilt angle is obtained by the ratio between horizontal and vertical mean gradients:

$$
\tan(\tau) = \frac{\langle d\log(v_i) \rangle}{dx_i} \bigg/ \frac{\langle d\log(v_i) \rangle}{dy_i}.
$$
(7.26)

Knowing the value of τ, the slant angle can be easily obtained for any point $\mathbf{x}_i$ from the mean gradient in the direction of tilt.

Particularly, at the point of fixation ($\mathbf{x}_i = 0$), we have:

$$
\tan(\sigma) = \frac{\langle d\log(v_i) \rangle}{d\mathbf{x}_\tau} \bigg|_{\mathbf{x}_i=0}.
$$
(7.27)

Knowing how to derive the local perspective information from the local spatial frequency gradients, we have yet to estimate this local frequency from the "cortical" filters' signals.

---

[2]This mean value of frequency over all orientations has the advantage of being independent of the local texture orientation, thus avoiding the risk of some aliasing.

7.5.3. *Mean Frequency from Log-Normal Filters*

Let us consider the ratio of the responses of two neighboring filters, at center frequencies f_i and f_{i+1}. After Knutsson *et al.* (1994), we have:

$$\frac{|G_{ri+1}(f)|^2}{|G_{ri}(f)|^2} = \exp\left(-\frac{1}{2\sigma_r^2}\left[\left(\ln\left(f/f_{i+1}\right)\right)^2 - \left(\ln\left(f/f_i\right)\right)^2\right]\right),$$

(7.28)

hence $|G_{ri+1}(f)|^2 = \left(f/\sqrt{f_i f_{i+1}}\right)^{((\ln(f_{i+1}/f_i))/\sigma_r^2)} |G_{ri}(f)|^2$, and if $\sigma_r^2 = \ln(f_{i+1}/f_i)$, a radial filter is deduced from the preceding one by the relation:

$$|G_{ri+1}(f)|^2 = \frac{f}{\sqrt{f_i f_{i+1}}} |G_{ri}(f)|^2.$$

(7.29)

As a consequence, it is possible to extract the mean (narrow-band) spatial frequency of the image in the vicinity of (f_i, f_{i+1}), by evaluating the ratio between two consecutive energy measures C_i and C_{i+1}:

$$\langle f \rangle_{i,j} = \frac{C_{i+1.}}{C_{i.}} = \frac{1}{\sqrt{f_i f_{i+1}}} \frac{\int_f f |G_{ri}(f)|^2 \int_\theta S(f,\theta) |G_{aj}(\theta)|^2 f d f d\theta}{\int_f |G_{ri}(f)|^2 \int_\theta S(f,\theta) |G_{aj}(\theta)|^2 f d f d\theta}.$$

Furthermore, in order to be independent of the local orientations, we can sum the C_{ij} coefficients over all orientations and obtain:

$$\langle f \rangle_i = \frac{1}{\sqrt{f_i f_{i+1}}} \frac{\int_f f |G_{ri}(f)|^2 \int_\theta S(f,\theta) \sum_j |G_{aj}(\theta)|^2 f d f d\theta}{\int_f |G_{ri}(f)|^2 \int_\theta S(f,\theta) \sum_j |G_{aj}(\theta)|^2 f d f d\theta}.$$

If we need more global information, a wide-band mean frequency can be obtained by the sum of the narrow-band mean frequencies $\langle f \rangle_i$, weighted by the relative energies of channels $C_{i.}$:

$$\langle f \rangle = \sum_i \frac{C_{i.}}{\sum_k C_{k.}} \langle f \rangle_i.$$

(7.30)

This provides a robust and generic measure of the mean local frequency, relatively independent of the local orientation/frequency shape, that is, for any local texture.

7.5.4. *Shape Recovery*

The image plane (Figure 7.15) is sampled by small overlapping sub-images. Each sub-image is filtered by the bank of Log-Normal filters, just as area V1 macro-columns do (7 orientations and 7 frequency bands).

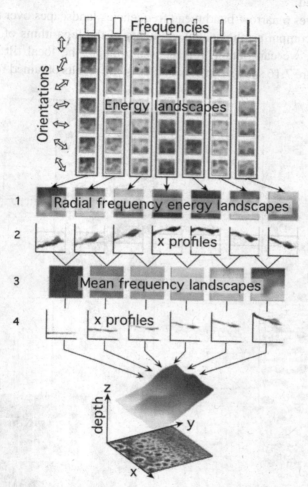

Figure 7.15. Principle of the estimation of perspective. Top matrix: the image is analyzed by 7 × 7 filters, giving 7 × 7 landscapes of energy. First line: the energies are summed over orientations, giving 7 circular energy landscapes, whose profiles are shown at line two. Line 3: the ratios between two consecutive landscapes gives the mean frequency landscapes, whose profiles appear in line four. Finally, the weighted sum of the mean frequencies leads to the depth information.

Figuring the complex cells responses, the output energy of each filter is computed, providing a set of 49 values for each sub-image. For each filter bank, the energies are summed over orientations, giving 7 energy profiles: one for each frequency band. Then the six ratios between neighboring bands are computed.

This gives 6 narrow-band mean frequencies landscapes over the whole image. By computing the spatial gradients of the logarithms of the mean frequencies, we can obtain, for each sub-image, the local tilt and slant values. Figure 7.16 shows some examples of the results obtained for various

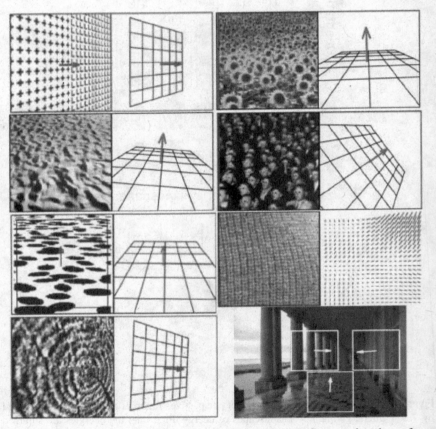

Figure 7.16. Some examples of perspective/shape estimation: the normal to the surfaces are accurately found for artificial textures (image #1) and similar to human perception for natural ones, where the true angles are not known (images #2 to 8). The local analysis of image 8 provides a good estimation of the orientations.

artificial or natural textures, whether regular or irregular. These results compare favorably with conventional computer vision algorithms on regular textures. It behaves fairly on non-stationary textures. However, in this case, very few comparisons can be made with computer vision.

7.6. Complementary Considerations

7.6.1. *Back to the Power Frequency Spectrum*

At the beginning if this chapter, we have insisted on the importance of the complex cells in V1. Knowing that they code the local energy spectrum of frequencies, we have shown the possible use of this model for scene categorization, for bottom-up attentional processes and for monocular perspective estimation. The main property of this coding is the fact that it is independent from small spatial translations of the image or of its content.

In fact, it has been demonstrated in psychophysics experiments that the human perception is mainly based on the analysis of the power spectrum of images (Guyader *et al.*, 2004). This has been successfully used in the design of artificial systems for image classification (Guyader *et al.*, 2002; Torralba and Oliva, 2003; Mermillod *et al.*, 2004; Mermillod *et al.*, 2005; Oliva, 2005).

7.6.2. *Small Number of Samples*

We have already said that for the frequency bands and orientations of the cortical filters, we choose an odd number of components. This fact is explained here with an example for a 1-D signal.

Let us suppose a continuous finite-length analog signal $i(x)$ of the continuous variable x, which can be sampled according to the Shannon theorem by a step Δx. Its sampled form is $i(k)$, k being discrete.

Let us now consider that the continuous signal is submitted to a translation: $i_t(x) = i(x - \xi)$, ξ not being an integer value of Δx. The shape of the sampled signal $i_t(k)$ is strongly modified, so that, in many cases, it is not recognizable anymore. How can we tell that $i(k)$ is the sampled version of the translated original? Let us take the Discrete Fourier Transform (DFT) of $i(k)$ over K samples:

$$I(n) = \sum_{k=0}^{K-1} i(k)\, e^{-j2\pi \frac{nk}{K}}. \tag{7.31}$$

With the translation ξ, it becomes:

$$I_t(n) = I(n) \, e^{-j2\pi \frac{n\xi}{K}}.$$

The inverse DFT of the sampled translated signal is:

$$i_t(k') = \sum_{k=0}^{K-1} I_t(n) \, e^{j2\pi \frac{nk'}{K}} \, e^{-j2\pi \frac{n\xi}{K}} = \sum_{k=0}^{K-1} i(k) \sum_{n=0}^{K-1} e^{j2\pi \frac{n(k'-k-\xi)}{K}}.$$

The sum of the last term is:

$$\sum_{n=0}^{K-1} e^{j2\pi \frac{n(k'-k-\xi)}{K}} = \frac{\sin \left[\pi \left(k' - k - \xi \right) \right]}{\sin \left[\pi \left(k' - k - \xi \right) / K \right]},$$

that is, an interpolation function so that the resulting discrete signal is:

$$i_t(k') = \sum_{k=0}^{K-1} i(k) \, \frac{\sin \left[\pi \left(k' - k - \xi \right) \right]}{\sin \left[\pi \left(k' - k - \xi \right) / K \right]}. \tag{7.32}$$

The convolution kernel of Equation (7.32) is an interpolation function, only when K is odd. Figure 7.17 gives the profile of the translation kernel for two cases: an odd and an even value of K, with the same translation value, $\xi = 0.8 \, \Delta x$.

It is clear that the odd value of K, which respects the periodicity of K, provides a suitable interpolation function.

In the case of image categorization, this coding allows the comparison of spectra between images rather independently of zoom or rotation effects, resulting in translations in the log-polar frequency plan (Guyader, 2004).

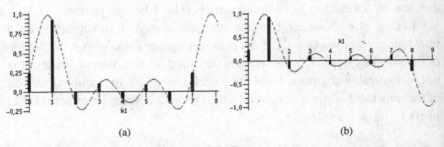

(a)   (b)

Figure 7.17. Interpolation kernel for signals subject to translations of a non-integer value of the sampling period. (a) odd number of samples (7). (b) even number of samples (8).

7.6.3. *Toward a Model of V4*

The seminal paper of Gallant *et al.* (1996) reported a study of cells in the V4 area. These cells exhibited a weak response to stimuli known to excite V1 cells, but responded very strongly to stimuli like concentric circles, figures with radii, and spiral shapes. The Fourier spectrum of these images also represents concentric circles, radii, and spirals.

In the log-polar frequency representation in V1 (Figure 7.4), these shapes become simply lines: concentric circles give vertical lines, radii give horizontal lines, and spirals give oblique lines. The simplest model for the cells described in Gallant *et al.* (1996) would be line detectors in the log-polar representation of spatial frequencies. A more in-depth study is given in (Kouh and Riesenhuber, 2003).

Another property of V4 cells appears as some scale invariance to the input patterns. There also is a recent model which has been proposed (Cadieu *et al.*, 2007).

A way to propose scale invariance is again to consider the log-polar representation: we have seen that in this representation, zooms and rotations will appear as simple translations. As the amplitude of the Fourier spectrum is invariant to translations, an interesting solution to the invariance problem is to take the Fourier power spectrum of the log-polar representation.

Of course, we can only speak of "relative" invariance. When translating an image, because of the limited field of vision, some objects disappear on one side and new ones appear on the opposite side. If the translation is too important, this may result in quite a new type of image. Similarly, when zooming an image, due to its limited frequency bandwidth, some low spatial frequencies disappear as some high frequencies appear. If the zoom is too high, this may also result in quite a new type of image. If you look at a tree through a telescope, you will not see the tree... all you will see are tree leaves.

Bibliography

Abbott, D, A Moini, A Yakovleff, XT Nguyen, R Beare, W Kim, A Bouzerdoum, RE Bogner and K Eshraghian (1996). Status of recent developments in collision avoidance using motion detectors based on insect vision. In *Proceedings of SPIE, Transportation Sensors and Controls: Collision Avoidance, Traffic Management, and ITS*, pp. 242–247.

Adam, G (1970). Electrical characteristics of ionic PSN junction as a model of the resting axon membrane. *Journal of Membrane Biology*, 3, pp. 291–312.

Adams, DL and JC Horton (2003). A precise retinotopic map of primate striate cortex generated from the representation of angioscotomas. *The Journal of Neuroscience*, 23(9), pp. 3771–3789.

Adams Jr JE (1998). Design of practical color filter array interpolation algorithms for digital cameras II. *Proceedings of the International Conference on Image Processing (ICIP'98)*. Chicago, IL, USA, 1, pp. 488–492.

Adelson EH and JR Bergen (1985). Spatio-temporal energy models for the perception of motion. *Journal of Optical Society of America*, 2(2), pp. 284–299.

Adelson EH and JA Movshon (1982). Phenomenal coherence of moving visual patterns. *Nature*, 300, pp. 523–525.

Adelson, EH and AP Pentland (1996). The perception of shading and reflectance. In *Visual Perception: Computation and Psychophysics*, Knill D and W Richards (Eds.), pp. 409–423. New York: Cambridge University Press.

Adelson, EH (1993). Perceptual organization and the judgment of brightness. *Science*, 262, pp. 2042–2044.

Aguilar M and WS Stiles (1954). Saturation of the rod mechanism of the retina at high levels of stimulation. *Optica Acta*, 1, pp. 59–65 (rods).

Albright, TD, R Desimone and CG Gross (1984). Columnar organization of directionally selective cells in visual area MT of the macaque. *Journal of Neurophysiology*, 51, pp. 16–31.

Aldroubi, A and H Feichtinger (1998). Exact iterative reconstruction algorithm for multivariate irregularly sampled functions In spline-like spaces: The l p-theory. *Proceedings of the American Mathematical Society*, 126(9), pp. 2677–2686.

Allan, LG and S Siegel (1997). Contingent color aftereffects: Reassessing old conclusions. *Perception and Psychophysics*, 59, pp. 129–141.

Alleysson, D (1999). *Le Traitement Chromatique dans la Rétine: Un Modèle de Base pour la perception Humaine des couleurs*. Doctoral dissertation, Université Joseph Fourier, Grenoble.

Alleysson, D and J Herault (2001). Variability in color discrimination data explained by a generic model with nonlinear and adaptive processing. *Color Research and Application*, 26(S1), pp. S225–S229.

Alleysson, D, S Suessttrunk and J Herault (2005). Linear demosaicing inspired by the human visual system. *IEEE Transactions on Image Processing*, 14(4), pp. 439–449.

Aloimonos J (1988). Shape from Texture, *Biological Cybernetics*, Vol. 58, pp. 345–360.

Anderson SJ and RF Hess (1990). Post-receptoral undersampling in normal human peripheral vision. *Vision Research*, 30(10), pp. 1507–1515.

Andreou, AG and KA Boahen (1995). A 590,000 transistor 48,000 pixel, contrast sensitive, edge enhancing, CMOS imager-silicon retina. In *Proceedings of the 16th Conference on Advanced Research in VLSI*, pp. 225–240.

Ans, B, C Marendaz, J Hérault and B Séré (2001). McCollough effect: A neural network model based on source separation. *Visual Cognition*, 8(6), pp. 823–841.

Arreguit, X, FA Van Schaik, FV Bauduin, M Bidiville and E Raeber (1996). A CMOS motion detector system for pointing devices. *IEEE Journal of Solid State Circuits*, 31(12), pp. 1916–1921.

Atchison D and G Smith (2000). *Optics of the Human Eye*. Butterworth-Oxford: Heinemann.

Atick J and A Redlich (1992). What does the retina know about natural scenes? *Neural Computation*, 4, pp. 196–210.

Atick, J, Z Li and A Redlich (1990). *Color Coding and its Interaction with Spatiotemporal Processing in the Retina.* Technical report IASSNS-HEP-90/75, NYU-NN-90/3.

Atick, JJ (1992). Could information theory provide an ecological theory of sensory processing. *Network*, 5, pp. 121–145.

Attwell, D, M Wilson and SM Wu (1984). A quantitative analysis of interactions between photoreceptors in the salamander (Ambystoma) retina. *Journal of Physiology*, 352, pp. 703–737.

Barlow HB, TP Kaushal and GJ Mitchison (1989). Finding minimum entropy codes. *Neural Computation*, 1, pp. 412–423.

Barlow, HB and WR Levick (1965). The mechanism of directionally selective units in the rabbit's retina, *Journal of Physiology*, 178, pp. 477–504.

Barlow, H and P Foldiac (1989). Adaptation and decorrelation in the cortex. In *The Computing Neuron*, R Durbin, C Miall and G Mitchison (Eds.), pp. 54–72. New York: Addison-Wesley.

Barlow, HB (1990). A theory about the functional role and synaptic mechanisms of visual after-effects. In *Vision: Coding and Efficiency*, C Blakemore (Ed.), pp. 363–375. New York: Cambridge University Press.

Barron, JL, DJ Fleet and SS Beauchemin (1994). Performances of optical flow techniques. *International Journal of Computer Vision*, 12(1), pp. 43–77.

Bartleson, CJ and F Grum (1984). *Optical Radiation Measurements*, Vol. 5. Visual Measurements. Orlando, FL: Academic Press.

Bayer, BE (1976). Color imaging array, US Patent 3,971,065, to Eastman Kodak Company, Patent and Trademark Office, Washington, DC, 1976.

Beaudot W, P Palagi and J Herault (1993). Realistic simulation tool for early visual processing including space, Time and Color Data. In *New Trends in Neural Computation*. J Mira, J Cabestany and A Prieto (Eds.), pp. 370–375. Springer-Verlag.

Beaudot, WHA (1994). *Le traitement Neuronal de l'Information dans la Rétine des Vertebrés* (in French). Ph.D thesis, INPG, Grenoble, France.

Beaudot, WHA (1995). An Adaptative Model with some Important Properties of Vertebrate Photoreceptors, In *Biophysics of Phototransduction: Molecular and Phototransductive Events*, C Taddei-Ferretti (Ed.), World Scientific Publishing, pp. 403–406.

Beaudot, WHA (1996). Sensory coding in the vertebrate retina: Towards an adaptive control of visual sensitivity. *Network: Computation in Neural Systems*, 7(2), pp. 317–323.

Beaudot, WHA and KT Mullen (2005). Orientation selectivity in luminance and color vision assessed using 2-D band-pass altered spatial noise. *Vision Research* 45, pp. 687–696.

Bell, AJ and TJ Sejnowski (1995). An information-maximization approach to blind separation and blind deconvolution. *Neural Computation*, 7, pp. 1129–1159.

Bellman, RE (1957). *Dynamic Programming*. Princeton, NJ: Princeton University Press.

Ben-Shahar O, PS Huggins, T Izo and SW Zucker (2003). Cortical connections and early visual function: Intra- and inter-columnar processing. *Journal of Physiology* — Paris 97, pp. 191–208.

Billock, VA (1995). Cortical simple cells can extract achromatic information from the multiplexed chromatic and achromatic signals in the parvocellular pathways. *Vision Research*, 35, pp. 2359–2369.

Black, MJ and R Rosenholtz (1995). Robust estimation of multiple surface shapes from occluded textures. *International Symposium on Computer Vision*, Miami, FL, pp. 485–490.

Bloomfield, SA (1995). A comparison of receptive field and tracer coupling size of horizontal cells in the rabbit retina. *Visual Neuroscience*, 12(5), 985–999.

Boahen, KA (1996). Retinomorphic vision systems. *IEEE Micro*, 16(5), pp. 30–39.

Bolduc, M and M Levine (1998). A review of biologically motivated space-variant data reduction models for robot vision. *Computer Vision and Image Understanding*, 69, pp. 170–184.

Bonmassar, G and EL Schwartz (1997). Space-variant Fourier analysis: The exponential chirp transform. *IEEE Transactions on Pattern Analysis and Machine Intelligence*, 19(10), pp. 1080–1089.

Bossomaier, TR, AW Snyder and A Hughes (1985). Irregularity and aliasing: Solution? *Vision Research*, 25(1), pp. 145–147.

Bouthemy, P and E François (1993). Motion Segmentation and qualitative Dynamic Scene Analysis from an Image Sequence. *International Journal of Computer Vision*, 10(2), pp. 157–182.

Bouvier, G, A Mahni and J Herault (1995). *A contrast and motion-sensitive silicon rétina*, ESSIRC-95.

Bouvier, G, A Mhani and G Sicard (1996). A contrast and motion sensitive silicon retina. *Proc. SPIE, Advanced Focal Plane Arrays and Electronic Cameras*, Berlin, 2950, pp. 131–136.

Boycott, BB, JM Hopkins and HG Sperling (1987). Cone connections of the horizontal cells of the rhesus monkey's retina. *Proceedings of the Royal Society B*, 229, pp. 345–379.

Boynton, RM (1979). *Human Color Vision*. New York: Holt, Rinehart and Winston.

Boynton, RM and J Gordon (1965). Bezold-Brücke hue shift measured by color-naming technique. *Journal of the Optical Society of America*, 55, pp. 78 –86.

Braddick, O (1993) Segmentation versus integration in visual motion processing. *Trends in Neuroscience*, 16, pp. 263–268.

Bradley, A, BC Scotun, I Ohzawa, G Sclar and RD Freeman (1987). Visual orientation and spatial frequency discrimination: A comparison of single neurons and behavior, *Journal of Neurophysiology*, 57, pp. 755–772.

Bradley, DC, N Qian and RA Andersen (1995). Integration of motion and stereopsis in cortical area MT of the macaque, *Nature*, 373, pp. 609–611.

Brainard, DH and BA Wandell (1992). Asymmetric color matching: How color appearance depends on the illuminant. *Journal of the Optical Society of America A, Optics and Image Science*, 9, pp. 1433–1448.

Brainard, DH, JM Kraft and P Longère (2003). Color constancy: Developing empirical tests of computational models. In *Color Perception: Mind and the Physical World*, R Mausfeld and D Heyer (Eds.), pp. 307–334. Oxford: Oxford University Press.

Brainard, D, P Longère, PB Delahunt, WT Freeman, JM Kraft and B Xiao (2006). Bayesian model of human color constancy. *Journal of Vision*, 6(11), pp. 1267–1281.

Brand P, R Mohr and Ph Bobet (1994). Distorsion optique: Correction dans un modèle projectif. In *Actes du 9ème Congrès AFCET de Reconnaissance des Formes et Intelligence Artificielle*, Paris, France, pp. 87–98.

Brelstaff, GJ, CA Parraga, T Troscianko and D Carr (1995). Hyper-spectral camera system: Acquistion and analysis, *SPIE Vol 2587* Geog. Inf. Sys. Photogram and Geolog./Geophys. Remote Sensing, pp. 150–159.

Brett, A, A Szmajda, U Grünert and PUR Martin. (2005). Mosaic properties of midget and parasol ganglion cells in the marmoset retina. *Visual Neuroscience*, 22, pp. 395–404.

Brown, SP and RH Masland (2001). Spatial scale and cellular substrate of contrast adaptation by retinal ganglion cells. *Nature Neuroscience*, 4, pp. 44–51.

Bruno, E and D Pellerin (2002). Robust motion estimation using spatial Gabor-like filters. *Signal Processing*, 82(2), pp. 175–187.

Budde, W (1983). *Optical Radiation Measurements*, Vol. 4. Physical Detectors of Optical Radiation. Orlando, FL: Academic Press.

Bullier, J (2001a). Feedback connections and conscious vision. *Trends in Cognitive Science*, 5, pp. 369–370.

Bullier, J (2001b). Integrated model of visual processing, *Brain Res. Brain Res. Rev.*, 36(2–3), pp. 96–107.

Bullier, J and JM Hupé (1996). Functional interactions between areas V1 and V2 in the Monkey. *Journal of Physiology*, 90, 112–119. (Paris)

Bullier, J, P Girard and PA Salin (1994). The role of area 17 in the transfer of information to extrastriate visual cortex. In *Primary Visual Cortex in Primates*. A Peters and KS Rockland (Eds.), Plenum Publishers 10, pp. 301–330.

Bülthoff, HH (1991). Shape from X: Psychophysics and computation, collection chapter. In *Computational Models of Visual Processing*, M Landy and A Movshon (Eds.), MIT Press, pp. 305–330.

Burnham, RW, RM Evans and SM Newhall (1957). Prediction of color appearance with different adaptation illuminations. *Journal of the Optical Society of America*, 47, pp. 35–42.

Buser, P and M Imbert (1992). *Vision*. Cambridge, MA: MIT Press.

Bushbaum, G and A Gottschalk (1983). Trichromacy, opponent colors coding and optimum color information transmission in the retina. *Proceedings of the Royal Society of London B*, 220, pp. 89–113.

Cadieu, C, M Kouh, A Pasupathy, CE Connor, M Riesenhuber and T Poggio (2007). A Model of V4 Shape Selectivity and Invariance. *Journal of Neurophysiology*, 98, pp. 1733–1750.

Cai, D, GC DeAngelis and RD Freeman (1997). Spatiotemporal receptive field organization in the lateral geniculate nucleus of cats and kittens. *Journal of Neurophysiology*, 78, pp. 1045–1061.

Calkins, DJ (2001). Seeing with S cones. *Prog Retin Eye Res*, 20(3), pp. 255–287.

Calkins, DJ and P Sterling (1996). Absence of spectrally specific lateral inputs to midget ganglion cells in primate retina. *Nature*, 381, pp. 613–615.

Calkins, DJ and P Sterling (1999). Evidence that circuits for spatial and color vision segregate at the first retinal synapse. *Neuron*, 24, pp. 313–321.

Calkins, DJ, Y Tsukamoto and P Sterling (1998). Microcircuitry and mosaic of a blue-yellow ganglion cell in the primate retina. *Journal of Neuroscience*, 18, pp. 3373–3385.

Callaway, EM (1998). Local circuits in primary visual cortex of the macaque monkey. *Ann Rev Neurosci*, 21, pp. 47–74.

Calow, D, N Krüger, F Wörgötter and M Lappe1 (2005). Biologically motivated space-variant filtering for robust optic flow processing. *Computation in Neural Systems*, 16(4), pp. 323–340.

Cappellini, V, A Mecocci and A Del Bimbo (1993). Motion analysis and representation in computer vision. *Journal of Circuits, Systems, and Computers*, 3(4), pp. 797–831.

Casagrande, VA (1999). The mystery of the visual K pathway. *Journal of Physiology*, 517(3), 630, pp. 907–917.

Castano-Moraga, CA, MA Rodriguez-Florido, L Alvarez, CF Westin and J Ruiz-Alzola (2004). Anisotropic interpolation of DT-MRI. In *MICCAI 2004, LNCS*, C Barillot, DR Haynor and P Hellier (Eds.), 3216, pp. 343–350.

Chander, D and EJ Chichilnisky (2001). Adaptation to temporal contrast in primate and salamander retina. *The Journal of Neuroscience*, 21(24), pp. 9904–9916.

Chapeau-Blondeau, F (1994). Information entropy maximization in the transmission by a neuron nonlinearity. *Comptes Rendus de L'Académie des Sciences, Paris*, Série II, 319, pp. 271–276.

Charkani, N and J Hérault (1995). On the performances of the fourth-order cross cumulants in blind separation of sources. *Paper presented at the ATHOS Workshop on Higher Order Statistics (IEEE Signal Processing)*, Begur, Spain.

Chauvin, A, J Hérault, C Marendaz and C Peyrin (2002). Natural scene perception: Visual attractors and images processing. In *Connectionist Models of Cognition and Perception*, JA Bullinaria and W Lowe (Eds.), pp. 236–248. Singapore: World Scientific Press.

Chen, TY, AC Bovik and LK Cormack (1999). Stereoscopic. ranging by matching image modulations. *IEEE Transactions on Image Processing*, 8(6), pp. 785–797.

Chichilnisky, EJ and BA Wandel (1999). Trichromatic opponent color classification. *Vision Research*, 39, pp. 3444–3458.

Chua, LO and L Yang (1988). Cellular neural naetworks. *IEEE Transactions on Circuits and Systems*, 35, pp. 1257–1290.

Cirrincione, G (1998). *A Neural Approach to the Structure from Motion Problem*. Ph.D thesis, INPG Grenoble, France.

Cirrincione, G, M Cirrincione and G Vitale (1994). Diagnosis of three-phase converters using the VQP neural network. *2nd IFAC Workshop on Computer Software Structures Integrating AI/KBS System in Process Control*, Lund, Sweden, 11/13 August 1994.

Cirrincione, G and M Cirrincione (1999). Robust neural approach for the estimation of the essential parameters in computer vision. *International Journal on Artificial Intelligence Tools*, 8(3), pp. 255–274.

Clerc, M and S Mallat (2002). The texture gradient equation for recovering shape from texture. *IEEE Transactions on Pattern Analysis and Machine Intelligence*, 24(4), pp. 536–549.

Cohen, JD (1975). Temporal independence of the Bezold-Brücke hue shift. *Vision Research*, 15, pp. 341–351.

Commission Internationale de l'Eclairage (1985). *Methods of Characterizing Illuminance Meters and Luminance Meters*. [Publication #69] CIE.

Comon, P (1994). Independent Component analysis: A new concept. *Signal Processing*, Elsevier, 36(3), pp. 287–314.

Comon, P and C Jutten (2007). *Separation de Sources, tome 1: Concepts de base et Analyse en Composantes Independantes*. Hermes, France.

Conway, BR, DH Hubel and MS Livingstone (2002). Color contrast in Macaque V1. *Cerebral Cortex*, 12, pp. 915–925.

Cowey, A and ET Rolls (1974). Human cortical magnification factor and its relation to visual acuity. *Experimental Brain Research*, 21(5), pp. 447–454.

Croner, LJ, K Purpura and E Kaplan (1993). Response variability in retinal ganglion cells of primates. *Proceedings of the National Academic Sciences of the USA*, 90, pp. 8128–8130.

Croner, LJ and E Kaplan (1995). Receptive fields of P and M ganglion cells across the primate retina. *Vision Research*, 35, pp. 7–24.

Curcio, CA and KA Allen (1990). Topography of ganglion cells in human retina. *Journal of Computational Neurology*, 1; 300(1), pp. 5–25.

Curcio, CA, KR Sloan, RE Kalina and KE Hendrickson (1990). Human photoreceptor topography. *Journal of Computational Neurology*, 292, pp. 497–523.

Dacey, DM, BB Lee, DK Stafford, J Pokorny and VC Smith (1996). Horizontal cells of the primate retina: Cone specificity without spectral opponency. *Science*, 2; 271(5249), pp. 656–659.

Dacey, DM and OS Packer (2003). Color coding in the primate retina: Diverse cell types and -specific circuitry. *Curr Opin Neurobiol*, 13(4), pp. 421–427.

Dacey, DM and MR Petersen (1992). Dendritic Field Size and Morphology of Midget and Parasol Ganglion Cells of the Human Retina. *PNAS*; 89, pp. 9666–9670.

Dacey, D, OS Packer, L Diller, D Brainard, B Peterson and BB Lee (2000). Center surround receptive field structure of cone bipolar cells in primate retina. *Vision Research*, 40, pp. 1801–1811.

Dacey, DM (1996). Circuitry for color coding in the primate retina. *Proceedings of the National Academic Sciences of the USA*, 93, pp. 582–588.

Dacey, DM (1999). The "blue-on" opponent pathway in primate retina originates from distinct bistratified ganglion cell type. *Nature*, 367, pp. 731–735.

Dacey, DM and BB Lee, (1994). Primate retina: Cell types, circuits and color opponency. *Progress in Retinal and Eye Research*, 18(6), pp. 737–763.

Dacheux, RF and E Raviola, (1982). Horizontal cells in the retina of the rabbit. *Journal of Neuroscience*, 2, pp. 1486–1494.

Dahmen, H, RW Wuest and J Zeil (1997). Extracting ego-motion parameters from optic flow: Principal limits for animals and machines. In *From Living Eyes to Seeing Machines*, MV Srinivasan and S Venkatesh (Eds.), pp. 174–198. Oxford, New York: Oxford University Press.

Das, A and CD Gilbert (1995). Long-range horizontal connections and their role in cortical reorganization revealed by optical recording of cat primary visual cortex, *Nature*, 375, pp. 780–784.

Daugman, JG (1980). Two-dimensional spectral analysis of cortical receptive field profiles. *Vision Research*, 20, pp. 847–856.

Daugman, JG (1992). Quadrature-phase simple-cell pairs are appropriately described in complex analytic form, *Journal of the Optical Society of America A*, 10(2), pp. 375–377.

De Monasterio, FM and P Gouras (1975). Functional properties of ganglion cells of the rhesus monkey retina. *Journal of Physiology*, 251, pp. 167–195.

De Valois, RL and K De Valois (1990). *Spatial Vision*. Oxford University Press.

De Valois, RL and KK De Valois (1993). A multi-stage color model. *Vision Research*, 33, pp. 1053–1065.

De Valois, RL, I Abramov and GH Jacobs, (1966). Analysis of response patterns of LGN cells. *Journal of the Optical Society of America*, 56, pp. 966–977.

DeAngelis, GC, I Ohzawa and RD Freeman (1993). Spatiotemporal organization of simple-cell receptive fields in the cat's striate cortex. I. General characteristics and postnatal development. *Journal of Neurophysiology*. 69(4), 1091–1117. II. Linearity of temporal and spatial summation. *Journal of Neurophysiology*, 69(4), pp. 1118–1135.

DeAngelis, GC, BG Cumming and WT Newsome (1998). Cortical area MT and the perception of stereoscopic depth. *Nature*, 394, 677–680.

Deco, G and B Schürmann (2001). Predictive coding in the visual cortex by a recurrent network with Gabor receptive fields. *Neural Processing Letters*, 14, pp. 107–114.

Delbrück, T (1993). Silicon retina with correlation-based velocity-tuned pixels. *IEEE Transactions on Neural Networks*, 4(3), pp. 529–541.

Demartines, P and J Herault (1997). Curvilinear component analysis: A self-organizing neural network for nonlinear mapping of data sets. *IEEE Transactions on Neural Networks*, 8(1), pp. 148–154.

Desimone, R and J Duncan, (1995). Neural mechanisms of selective visual attention. *Annual Revue of Neuroscience*, 18, pp. 193–222.

Desimone, R and SJ Schein (1987). Visual properties of neurons in area V4 of the macaque: Sensitivity to stimulus form. *Journal of Neurophysiology*, 57, pp. 835–868.

Deutschmann, RA and C Koch (1998). Compact real-time 2-D gradient-based analog VLSI motionsensor. In *Advanced Focal Plane Arrays and Electronic Cameras II*, TM Bernard (Ed.), Proc. SPIE, Vol. 3410, pp. 98–108.

Deutschmann, RA, CM Higgins and C Koch, (1997). Real-time analog VLSI sensors for 2-D direction of motion. In *Proc. International Conference on Artificial Neural Networks (ICANN97)*, W Gerstner, A Germound, M Hasler and JD Nicoud (Eds.), *volume 1327 Lecture Notes in Computer Science*, pp. 1163–1168, Lausanne, Switzerland, October 1997. Springer Verlag.

DeValois, RL (1991). Orientation and spatial frequency selectivity: Properties and modular organization. In *From Pigments to Perception*, A Valberg and BB Lee (Eds.). New York: Plenum.

DeValois, RL, DG Albrecht and LG Thorell (1982). Spatial frequency selectivity of cells in macaque visual cortex. *Vision Research*, 22, pp. 545–559.

Ding, Y and VA Casagrande (1997). The distribution and morphology of LGN K pathway axons within the layers and CO blobs of owl monkey V1. *Visual Neuroscience*, 14, pp. 691–704.

Djamgoz, MB and H Kolb (1993). Ultrastructural and functional connectivity of intracellularly stained neurones in the vertebrate retina: Correlative analyses. *Microscopy Research and Technique*, 1, 24(1), pp. 43–66.

Dong, DW and JJ Atick (1995a). Statistics of natural time varying images. *Network Computation in Neural Systems*, 6(3), pp. 345–358.

Dong, DW and JJ Atick (1995b). Temporal decorrelation: A theory of lagged and nonlagged responses in the lateral geniculate nucleus, *Network: Computations in Neural Systems*, 6, pp. 159–178.

Donoho, DL, M Vetterli, RA DeVore and I Daubechies (1998). Data compression and harmonic analysis. *IEEE Transaction on Information Theory*, 44(6), pp. 2435–2476.

Dow, BM, JD Snyder, RG Vautin and R Bauer (1981). Magnification factor and receptive field size in foveal striate cortex of the monkey. *Experimental Brain Research*, 44, pp. 89–97.

Dowling, JE (1987). *The Retina: An Approachable Part of the Brain.* Cambridge, MA: Harvard University Press.

Drasdo, N and CW Fowler (1974). Non-linear projection of the retinal image in a wide-angle schematic eye. *British Journal of Ophthalmology*, 58(8), pp. 709–714.

Drew, PJ and LF Abbott (2006). Models and properties of power-law adaptation in neural systems. *Journal of Neurophysiology*, 96, pp. 826–833.

Duffy, CJ and RH Wurtz (1991). Sensitivity of MST neurons to optic flow stimuli. I. A continuum of response selectivity to large-field stimuli, *Journal of Neurophysiology*, 65, pp. 1329–1345.

Duhamel, JR, CL Colby and ME Goldberg (1992). The updating of the representation of visual space in parietal cortex by intended eye-movements, *Science*, 255, pp. 90–92.

Duncan, RO and GM Boynton (2003). Cortical magnification within human primary visual cortexcorrelates with acuity thresholds. *Neuron*, 38(4), pp. 659–671.

Dupont, P, GA Orban, B De Bruyn, A Verbruggen and L Mortelmans (1994). Many areas in the human brain respond to visual motion. *Journal of Neurophysiology*, 72, pp. 142–1424.

D'Zmura, M and G Iverson (1993). Color constancy. I. Basic theory of two-stage linear recovery of spectral descriptions for lights and surfaces. *Journal of the Optical Society of America A*, 10, pp. 2148–2165.

Ejima, Y and S Takahashi (1984). Bezold-Brücke hue shift and nonlinearity in opponent-color process. *Vision Research*, 24, pp. 1897–1904.

Etienne-Cummings, RR, J Van Der Spiegel and P Mueller (1997). A focal plane visual motion measurement sensor. *IEEE Transactions on Circuits and Systems I: Fundamental Theory and Applications*, 44(1), pp. 55–66.

Fahey, PK, DA Burkhardt (2003). Center-surround organization in bipolar cells: Symmetry for opposing contrasts. *Visual Neuroscience*, 20(1), pp. 1–10.

Fahle, M and T Poggio (1981). Visual hyperacuity: spatio-temporal interpolation in human vision. *Proceedings of the Royal Society of London B*, 213, pp. 1783–1796.

Fairchild, MD (1998). *Color Appearance Models*. Addison-Wesley, Reading, Mass. ISBN 0-201-63464-3.

Faugeras, O (1993). *Three-Dimensional Computer Vision — A Geometric Viewpoint*. Artificial intelligence. Cambridge, MA: MIT Press.

Felleman, DJ and DC Van Essen (1991). Distributed hierarchical processing in the primate cerebral cortex. *Cerebral Cortex*, 1, pp. 1–47.

Field, DJ (1994). What is the goal of sensory coding? *Neural Computation*, 6, pp. 559–601.

Field, DJ (1987). Relations between the statistics of natural images and the response properties of cortical cells. *Journal of the Optical Society of America A*, 4, pp. 2379–2394.

Finkelstein, MA, M Harrison and DC Hood (1988). Receptoral and post-receptoral non-linearities and foveal increment thresholds. *Investigative Ophtalmology and Visual Science*, 29(suppl.), 300.

Fischer, S, F Šroubek, L Perrinet, R Redondo and G Cristóbal, (2007). Self-invertible 2D Log-Gabor wavelets. *International Journal of Computer Vision*, 75(2), pp. 231–246

Franceschini, N (1998). Combined optical, neuroanatomical, electro-physiological and behavioural studies on signal processing in the fly compound eye. In *Biocybernetics of Vision: Integrative Mechanisms and Cognitive Processes*, C Taddei-Ferretti (Ed.), pp. 341–361. Singapore: World Scientific.

Franceschini, N, JM Pichon and C Blanes, (1992). From insect vision to robot vision, *Philosophy Transactions of the Royal Society B*, 337, pp. 283–294.

Freed, MA, R Pflug, H Kolb and R Nelson (1995). ON-OFF amacrine cells in cat retina. *Journal of Comparative Neurology*, 364, pp. 556–566.

Freeman, AW (1991). Spatial characteristics of the contrast gain control in the cat's retina. *Vision Research*, 31(5), pp. 775–785.

Friedman, JH (1987). Exploratory projection pursuit, *Journal of the American Statistical Association*, 82(397), pp. 249–266.

Frisby, JP (1979). *Seeing: Illusion, Brain and Mind.* Walton Street, Oxford: Oxford University Press.

Funatsu, E, Y Nitta, Y Miyake, T Toyoda, K Hara, H Yagi, J Ohta and K Kyuma (1995). An artificial retina chip with a 256×256 array of n-MOS variable sensitivity photodetector cells. *Proc. SPIE, Machine Vision Applications, Architectures, and Systems Integration IV*, 2597, pp. 283–291.

Gabor, D (1946). Theory of communication. *Journal of Institute of Electrical Engineers*, 93, pp. 429–457.

Gallant, JL, CE Connor, S Rakshit, JW Lewis and DC Van Essen (1996). Neural responses to polar, hyperbolic, and Cartesian gratings in area V4 of the macaque monkey. *Journal of Neurophysiology*, 76, pp. 2718–2739.

Galvin, SJ, RP O'Shea, AM Squire and DG Govan (1997). Sharpness overconstancy in peripheral vision. *Vision Research*, 37, pp. 2035–2039.

Gårding, J (1993). Direct estimation of shape from texture, *IEEE Transactions on Pattern Analysis and Machine Intelligence*, 15(11), pp. 1202–1208.

Gårding, J and T Lindeberg (1996). Direct computation of shape cues using scale-adapted spatial derivative operators. *International Journal of Computer Vision*, 17(2), pp. 163–191.

Geisler, WS and DB Hamilton (1986). Sampling-theory analysis of spatial vision. *Journal of the Optical Society of America A*, 3(1), pp. 62–70.

Gibson, J (1950). *The Perception of the Visual World.* Boston: Houghton Mifflin.

Glotzbach, JW, RW Schafer and K Illgner (2001). A method of color filter array interpolation with alias cancellation properties. *Proc. Int'l Conf. on Image Processing (ICIP'01)*, Thessaloniki, Greece, 1, pp. 141–144.

Gonzalez, RC and RE Woods (1992). *Digital Image Processing*. Addison-Wesley Publishing Company.

Goodchild, AK, KK Ghosh and PR Martin (1996). Comparison of photoreceptor spatial density and ganglion cell morphology in the retina of human, macaque monkey, cat, and the marmoset callithrix jacchus. *The Journal of Comparative Neurology*, 366, pp. 55–75.

Gottlieb, J, M Kusunoki and ME Goldberg (1998). The representation of visual salience in monkey parietal cortex. *Nature*, 391, pp. 481–484.

Gouras, P (2003). The role of S-cones in human vision. *Doc. Ophthalmol.*, 106(1), pp. 5–11.

Granger, EM and JC Hurtley (1973). Visual chromaticity modulation transfer function. *Journal of the Optical Society of America*, 63, pp. 1173–1174.

Grum, F and CJ Bartleson (1980). *Optical Radiation Measurements*, Vol. 2. Color Measurement. New York: Academic Press.

Guérin-Dugué, A and M Elghadi (1999). Shape from Texture by Local Frequencies Estimation, *SCIA99*, pp. 533–544, June 7–11, Kangerlussuaq, Greenland.

Guérin-Dugué, A and A Oliva (2000). Classification of scene photographs from local orientations features. *Pattern Recognition Letters*. 21, pp. 1135–1140.

Guérin-Dugué, A and PM Palagi (1994). Texture segmentation using pyramidal gabor function and self-organizing feature maps, *Neural Processing Letters*, 1(1), pp. 25–29.

Guérin-Dugué, A, P Teissier, G Delso Gafaro and J Herault (1999). Curvilinear Component Analysis for high dimensional data representation: II. Examples of additional mapping constraints in specific applications. In *Proceedings of IWANN'99*, J Mira, JV Sanchez (Eds.), Vol. II, Springer, Alicante, Spain, pp. 635–644.

Guth, SL and HR Lodge (1973). Heterochromatic additivity, foveal spectral sensitivity and a new color model. *Journal of Optical Society of America*, 63, pp. 450–462.

Guyader, N (2004). *Perception visuelle et analyse des scènes: Modèles de structures corticales*. Ph.D thesis, Université Joseph Fourier, Grenoble, France.

Guyader, N, A Chauvin, C Peyrin, J Hérault and C Marendaz (2004). Image phase or amplitude? Rapid scene categorization is an amplitude-based process. *C. R. Biologies*, 327, pp. 313–318.

Guyader, N, H Le Borgne, J Hérault and A Guérin-Dugué (2002). Towards the introduction of human perception in a natural scene classification system. *Proc. of the IEEE Int. workshop on Neural Networks for Signal Processing (NNSP'2002)*, pp. 385–394, Martigny Valais, Switzerland.

Hahn, LW and WS Geisler (1995). Adaptation mechanisms in spatial vision-II Flash thresholds and background adaptation. *Vision Research*, 35(44), pp. 1595–1609.

Hamer, RD and CW Tyler (1995). Phototransduction: Modeling the primate cone flash response. *Visual Neuroscience*, 12(6), 1063–1082.

Hare, WA and WG Owen (1990) Spatial organization of the bipolar cell's receptive field in the retina of the tiger salamander. *Journal of Physiology*, 421, pp. 223–245.

Hare, WA and WG Owen (1992). Effect of 2-amino-4-phosphonobutyric acid on cells in the distal layers of the tiger salamander retina. *Journal of Physiology*, 445, pp. 741–757.

Hartley, R, R Gupta and T Chang (1992). Stereo from uncalibrated cameras. In *Proceedings of the Conference on Computer Vision and Pattern Recognition*, Urbana-Champaign, Illinois, USA, pp. 761–764.

Harvey, LO and VV Doan (1990). Visual masking at different polar angles in the two-dimensional Fourier plane. *Journal of Optical Society of America A*, 7, pp. 116–127.

Havlicek, JP and AC Bovik (1994). Multi-Component AM-FM Image Models and Wavelets based Demodulation with Components Tracking, *IEEE International Conference on Image processing*, Austin, USA, pp. I.41–I.45.

Havlicek, JP and AC Bovik (1995). *AM–FM Models, the Analytic Image, and Nonlinear Demodulation Techniques*, Technical report, CVIS-TR-95-001.

Hawken, MJ and AJ Parker (1987). Spatial Properties of Neurons in the Monkey Striate Cortex. *Proceedings of the Royal Society of London, Series B, Biol Sci*, 231(1263), pp. 251–288.

Hebel, R and H Holländer (1983). Size and distribution of ganglion cells in the human retina. *Anatomy and Embryology*, 168(1), pp. 125–136.

Heeger, DJ (1987) Model for the extraction of image flow. *Journal of Optical Society of America*, 4(8), pp. 1455–1471.

Heeger, D, E Simoncelli and J Movshon (1996). Computational models of cortical visual processing. *Proceedings of the National Academic Sciences of the USA*, 93, 623–627.

Held, R and SR Shattuck (1971). Color and edge-sensitive channels in the human visual system: Tuning for orientation. *Science*, 174, pp. 314–316.

Henderson, A and JM Hollingworth (1998). Does consistent scene context facilitate object perception. *Journal of Experimental Psychology: General*, 127(4), pp. 398–415.

Hendry, SHC and TY Yoshioka (1994). A neurochemically distinct third channel in the macaque dorsal lateral geniculate nucleus. *Science*, 264, pp. 575–577.

Hérault, J (1996). A model of color processing in the retina of vertebrates: From photoreceptors to color opposition and color constancy. *Neurocomputing*, 12(2–3), pp. 113–129.

Hérault, J and B Ans (1984). Réseau de neurones à synapses modifiables: Décodage de messages sensoriels composites par apprentissage non supervisé et permanent. *Compte Rendus de l'Académie des Sciences, Paris*, Série III, 299, pp. 525–528.

Hérault, J and C Jutten (1986). Space or time adaptive signal processing by neural networks models. *AIP Conf. Proc*, 151, pp. 206–211.

Hérault, J, C Jaussions-Picaud and A Guérin-Dugué (1999). Curvilinear component analysis for high dimensional data representation: I. Theoretical aspects and practical use in the presence of noise. In *Proceedings of IWANN'99*, J Mira and JV Sanchez (Eds.), Vol. II, Springer, Alicante, Spain, pp. 625–634.

Hérault, J, A Oliva, A Guérin-Dugué (1997). Scene categorisation by curvilinear component analysis of low frequency spectra. *European Symposium on Artificial Neural Networks*, Bruges, BE.

Higgins, CM and C Koch (1997). Analog CMOS velocity sensors. In *Electronic Imagin'97*, San Jose, CA, February 1997.

Hild, M (1990). An algorithmic approach to modelling the retinal receptor topography. *Biological Cybernetics*, 62, pp. 511–518.

Hodgkin, AL and H Huxley (1952). A quantitative description of membrane current and its application to conduction and excitation in nerve. *Journal of Physiology (London)*, 128, pp. 500–544.

Horn, BKP (1986). *Robot Vision*. Cambridge, MA: MIT Press.

Horn, BKP and BG Schunk (1981). Determining Optical Flow. *Artificial Intelligence*, 17, pp. 185–203.

Hotelling, H (1933). Analysis of a complex of statistical variables into principal components, *Journal of Educational Psychology*, 24; 417–441, pp. 498–520.

Hsu, A, RG Smith, G Buchsbaum and P Sterling (2000). Cost of cone coupling to trichromacy in primate fovea. *J. Opt. Soc. Am. A*, 17(3), pp. 635–640.

Hubel, DH and TN Wiesel (1962). Receptive fields, binocular interaction, and functional architecture in the cat's visual cortex. *Journal of Physiology*, 160, pp. 106–154.

Hubel, DH and TN Wiesel (1963). Receptive fields of cells in striate cortex of very young, visually inexperienced kittens. *Journal of Physiology*, 26, pp. 994–1002.

Hubel, DH and TN Wiesel (1968). Receptive fields and functional architecture of monkey striate cortex. *Journal of Physiology*, 195, pp. 215–243.

Hubel, DH and TN Wiesel (1974). Sequence regularity and geometry of orientation columns in the monkey striate cortex. *Journal of Computational Neurology*, 158, pp. 267–294.

Hurlbert, A (1998). Computational models of color constancy. In *Perceptual Constancies: Why Things Look as they Do*, V Walsh and J Kulikowski (Eds.), pp. 283–322. Cambridge, UK: Cambridge University Press.

Hwang, WS, CS Lu and PC Chung (1998). Shape from texture estimation of planar surface orientation through the ridge surfaces of continuous wavelets transform. *IEEE Transactions on Image Processing*, 7(5), pp. 773–780.

Hyvärinen, A, J Karhunen and E Oja (2001). *Independent Component Analysis*. New York: Wiley.

Imho, SM, VJ Volbrecht and JL Nerger (2004). A new look at the Bezold-Brücke hue shift in the peripheral retina. *Vision Research*, 44, pp. 1–16.

Itti, L, C Koch and E Niebur (1998). Model of saliency-based visual attention for rapid scene analysis. *IEEE Transactions on Pattern Analysis and Machine Intelligence*, 20, pp. 1254–1259.

Jacobs, A L and FS Werblin (1998). Spatiotemporal patterns at the retinal output. *Journal of Neurophysiology*, 80(1), pp. 447–451.

Jacobs, A, T Roska and F Werblin (1996). Methods for constructing physiologically motivated neuromorphic models in CNNs. *International Journal of Circuit Theory Applications*, 24, pp. 315–339.

Jacobs, GH (1996). Primate photopigments and primate color vision. *Proceedings of the National Academy of Sciences of USA*, 93, pp. 577–581.

Jameson, D and LM Hurvich (1955). Some quantitative aspects of an opponent-colors theory: I. Chromatic responses and spectral saturation. *Journal of the Optical Society of America*, 45, pp. 546–552.

Johnson, EN, MJ Hawken and R Shapley (2001). The spatial transformation of color in the primary visual cortex of the macaque monkey. *Nature Neuroscience*, 4(4), pp. 409–416.

Jolliffe, IT (1986). *Principal Component Analysis*, New York: Springer.

Jonas, JB, U Schneider and GOH Naumann (1992). Count and density of human retinal photoreceptors. *Graefe's Arch. Clin. Exp. Ophthalmol.*, 230, pp. 505–510.

Jones, JP and LA Palmer (1987). An evaluation of the twodimensional Gabor filter model of simple receptive fields in cat striate cortex. *Journal of Neurophysiology*, 58, pp. 1233–1258.

Jones, PD and DH Holding (1975). Extremely long-term persistence of the McCollough effect. *Journal of Experimental Psychology: Human Perception and Performance*, 1, pp. 323–327.

Jutten, C and P Comon (2007). *Separation de Sources, tome 2: Au dela de l'aveugle, et Applications*. France: Hermes.

Jutten, C and J Hérault (1991). Blind Separation of sources. Part. I: An adaptive algorithm based on neuromimetic architecture. *Signal Processing*, 24, pp. 1–10.

Kaiser, PK and RM Boynton (1996). *Human Color Vision*, Second Edition: Washington, DC, Optical Society of America.

Kambhatla, N and TK Leen (1994). Dimension reduction by local principal component analysis, *Neural Computation*, 9(7), pp. 1493–1516.

Kamermans, M and H Spekreise (1999). The feedback pathway from horizontal cells to cones, a mini review with a look ahead. *Vision Research*, 39, pp. 2449–2468.

Kamermans, M, DA Kraaji and H Spekreise (1998). The cone/horizontal cell system: A possible site for color constancy. *Visual Neuroscience*, 15, pp. 787–797.

Kanatani, K (1993). *Geometric Computation for Machine Vision*. Oxford Science Publications: Oxford.

Kanatini, K and T Chou (1989). Shape from texture: General principle, *Artificial Intelligence*, 38, pp. 1–48.

Kandel, ER, JH Schwartz and TM Jessell (Eds.) (1991). *Principles of Neural Science*, 3rd ed., New York: Elsevier Science Publishing Co.

Kaneko, A and M Tachibana (1986). Effects of gamma-aminobutyric acid on isolated cone photoreceptors of the turtle retina. *Journal of Physiology*, 373, pp. 443–461.

Kanizsa, G (1979). *Organization in Vision*. New York: Praeger.

Kelly, DH (1975). Luminance and chromatic flickering patterns have opposite effects. *Science*, 1988, pp. 371–372.

Kim, KJ and F Rieke (2001). Temporal contrast adaptation in the input and output signals of salamander retinal ganglion cells. *Journal of Neuroscience*, 21, pp. 287–299.

Kingdom, AA and KT Mullen (1995). Separating color and luminance information in the visual system. *Spatial Vision*, 9(2), pp. 191–219.

Kingslake, R (1965). *Applied Optics and Optical Engineering*. New York: Academic Press.

Knill, DC (1995). Discriminating surface slant from texture: Comparing human and ideal observers. *Vision Research*, 38(11), pp. 1683–1711.

Knutsson, H and M Andersson (2003). Loglets — Generalized Quadrature and Phase for Local Spatio-temporal Structure Estimation. *Scandinavian Conference on Image Analysis* (SCIA 2003).

Knutsson, H, CF Westin and G Granlund (1994). Local multiscale frequency and bandwidth estimation. *IEEE International Conference on Image Processing* (*ICIP'94*), Austin, Texas.

Kohn, A (2007). Visual adaptation: Physiology, mechanisms, and functional benefits. *Journal of Neurophysiology*, 97, pp. 3155–3164.

Kolb, H and RW West (1977). Synaptic connections of the interplexiform cell in the retina of the cat. *Journal of Neurocytology*, 6, pp. 155–170.

Kolb, H, KA Linberg and SK Fisher (1992). Neurons of the human retina: A Golgi study. *Journal of Comparative Neurology*, 318, pp. 147–187.

Kolb, H, A Mariani and A Gallego (1980). A second type of horizontal cell in the monkey retina. *Journal of Comparative Neurology*, 189, pp. 31–44.

Kouh, M and M Riesenhuber (2003). *Investigating Shape Representation in Area V4 with HMAX: Orientation and Grating Selectivities*. AI Memo 2003-021, September 2003. MIT, Cambridge, MA, USA.

Kramer, J (1996). Compact integrated motion sensor with three-pixel interaction. *IEEE Transactions on Pattern Analysis and Machine Intelligence*, 18(4), pp. 455–460.

Krizaj, D and DR Copenhagen (2002). Calcium regulation in photoreceptors, *Frontiers in Bioscience*, 7, pp. d2023–d2044.

Kulikowski, J, S Marcelja and PO Bishop (1982). Theory of spatial position and spatial frequency relations in the receptive fields of simple cells in the visual cortex. *Biological Cybernetics*, 43, pp. 187–198.

Kunken, JM, H Sun and BB Lee (2005). Macaque ganglion cells, light adaptation, and the Westheimer paradigm. *Vision Research*, 45, pp. 329–341.

Lamme, VAF and PR Roelfsema (2000). The distinct modes of vision offered by feedforward and recurrent processing. *Trends in Neurosciences*, 23, pp. 571–579.

Land, EH and JJ McCann (1971). Lightness and retinex theory. *Journal of the Optical Society of America*, 61, pp. 1–11.

Laurent, G and F Gabbiani (1998). Collision-avoidance: Nature's many solutions. *Nature neuroscience*, 1(4), pp. 261–263.

Le Borgne, H, N Guyader, A Guérin-Dugué and J Hérault (2003). Classification of images: ICA filters Vs Human Perception. *Proceedings of the 7th International Symposium on Signal Processing and Its Applications* (IEEE catalog N°03EX714C, ISBN 0-7803-7947-0), (ISSPA 2003), vol 2, pp. 251–254, Paris, France, July 1–4 2003.

Lee, BB, DM Dacey, VC Smith and J Pokorny (2003). Dynamics of sensitivity regulation in primate outer retina: The horizontal cell network. *Journal of Vision*, 3(7), pp. 513–526.

Lee, BB, J Kremers and T Yeh (1998). Receptive fields of primate retinal ganglion cells studied with a novel technique. *Visual Neuroscience*, 15, pp. 161–175.

Lee, JA, A Lendasse and M Verleysen (2004). Nonlinear projection with curvilinear distances: Isomap versus curvilinear distance analysis. *Neurocomputing*, 57, pp. 49–76.

LeGrand, Y (1970). *Light, Color, and Vision*. New York: John Wiley and Sons.

Leipnik, RB (1991). On the Log-Normal random variables: I- The characteristic function. *Journal of Australian Mathematical Society, Series B*, 32, pp. 327–347.

Lelandais, S, L Boutté and J Plantier (2005). Shape from texture: Local scales and vanishing line computation to improve results for macrotextures. *International Journal of Image Graphics*, 5(2), pp. 329–350.

Leventhal, AG, YC Wang, MT Schmolesky and Y Zhou (1998). Neural correlates of boundary perception. *Visual Neuroscience*, 15, pp. 1107–1118.

Li, A and Q Zaidi (2004). Three-dimensional shape from non-homogeneous textures: carved and stretched surfaces. *Journal of Vision*, 4(10(3)), pp. 860–878.

Liu, Y, R Vogels and GA Orban (2004). Convergence of depth from texture and depth from disparity in macaque inferior temporal cortex. *The Journal of Neuroscience*, 24(15), pp. 3795–3800.

Loh, A and P Kovesi (2005). *Shape from Texture without Estimating Transformations*. Technical report, UWA-CSSE-05-001, July 2005.

Long, KO and SK Fisher (1983). The distributions of photoreceptors and ganglion cells in the California ground squirrel, spermophilus beecheyi. *The Journal of Comparative Neurology*, 221, pp. 329–340.

Luong, QT and O Faugeras (1996). The fundamental matrix: Theory, algorithms and stability analysis. *International Journal of Computer Vision*, 17(1), pp. 43–76.

Luthon, F and D Dragomirescu (1999). A cellular analog network for MRF-based video motion detection. *IEEE Transactions on Circuits and Systems. I: Fundamental Theory and Applications*, 46(2), pp. 281–293.

MacAdam, DL (1942). Visual sensitivities to color differences in daylight. *Journal of the Optical Society of America*, 32, pp. 247–273.

Mahowald, M (1994). *An Analog VLSI System for Stereoscopic Vision*. UK: Kluwer Academic Publishers.

Malik, J and R Rosenholtz (1997). Computing local surface orientation and shape from texture for curved surfaces. *International Journal of Computer Vision*, 23(2), pp. 149–168.

Maloney, LT and JN Yang (2001). The illuminant estimation hypothesis and surface color perception. In *Color Perception: From Light to Object*, R Mausfeld and D Heyer (Eds.), pp. 335–358. Oxford: Oxford University Press.

Maloney, LT (1996). Some implications for retinal sampling and reconstruction. *Exploratory Vision: The Active Eye* (Springer series in perception engineering) (Chapter 5), MS Landy, LT Maloney and M Pavel (Eds.), pp. 121–156. New York: Springer Verlag.

Mamassian, P, MS Landy and LT Maloney (2002). Bayesian modelling of visual perception. In *Probabilistic Models of the Brain: Perception and Neural Function*, RPN Rao, BA Olshausen and MS Lewicki (Eds.), pp. 13–36. Cambridge, MA: MIT Press.

Marc, RE and HG Sperling (1977). Chromatic organization of primate cones. *Science*, 196(4288), pp. 454–456.

Marcelja, S (1980). Mathematical description of the responses of simple cortical cells. *Journal of Optical Society of America*, 70(11), pp. 1297–1300.

Marendaz, C, A Chauvin and J Hérault (2005). A causal link between scene exploration, local saliency and scene context [Abstract]. *Journal of Vision*, 5(8), 919, 919a, http://journalofvision.org/5/8/919/, doi:10.1167/5.8.919.

Marimont, DH and BA Wandel (1992). Linear models of surface and illuminant spectra. *Journal of Optical Society of America A*, 9, pp. 1905–1913.

Marr, D (1982). *Vision*. Freeman.

Marr, D and E Hildreth (1980). Theory of edge detection. *Proceedings of the Royal Society of London*, Series B, Biol. Sci., 207(1167), pp. 187–217.

Martin, PR and U Grünert (2003). Ganglion Cells in Mammalian Retinae. In *The Visual Neurosciences*, LM Chalupa and JS Werner (Eds.), pp. 410–421. Cambridge, MA: The MIT press.

Martin, PR (1998). Color processing in the primate retina: Recent progress. *Journal of Physiology*, 513(3), pp. 631–638.

Martinez-Uriegas, E (1990). Spatiotemporal multiplexing of chromatic and achromatic information in human vision. *Human Vision and Electronic Imaging: Models, Methods and Applications*, SPIE, 1249.

Massot, C and J Hérault (2008). Model of Frequency Analysis in the Visual Cortex and the Shape from Texture Problem. *Int. Journal of Computer Vision*, 76, pp. 165–182.

Mastronarde, DN (1987). Two classes of single-input X-cells in cat lateral geniculate nucleus. I. Receptive field properties and classification of cells. *Journal of Neurophysiology*, 57, pp. 357–380.

Matthias, O, MO Franz and HG Krapp (1998). Wide-Field, Motion-Sensitive Neurons and Optimal Matched Filters for Optic Flow. Technical report no. 61, Max-Planck-Institut fuer biologische Kybernetik, Tuebingen, Germany.

McCollough, C (1965). Color adaptation of edge-detectors in the human visual system. *Science*, 149, pp. 1115–1116.

McIlwain, JT (1996). *An Introduction to the Biology of Vision*. Cambridge, UK: Cambridge University Press.

McNaughton, PA (1993). The light response of photoreceptors. In *VISION: Coding and Efficiency*, C. Blackmore (Ed.), Cambridge University Press.

Mead, C And MA Mahowald (1988). A silicon model of early visual processing. *Neural Networks*, 1, pp. 91–97.

Mermillod, M, N Guyader and A Chauvin (2004). Does the energy spectrum from Gabor wavelet filtering represent sufficient information for neural network recognition and classification tasks? In *Connectionist Models of Cognition, Perception and Emotion*, H Bowman and C Labiouse (Eds.), pp. 148–156. Singapore: World Scientific.

Mermillod, M, N Guyader and A Chauvin (2005). The coarse-to-fine hypothesis revisited: Evidence from neuro-computational modeling. *Brain and Cognition*, 57(2), pp. 151–157.

Meylan, L, D Alleysson and S Süsstrunk (2007). A model of retinal local adaptation for the tone mapping of color filter array images, *Journal of Optical Society of America A*, 24(9), pp. 2807–2816.

Mhani, A, G Sicard and G Bouvier (1997). Analog vision chip for sensing edges contrasts and motion. *ISCAS*, pp. 326–329.

Miyoshi, A, I Mitsuo and N Takehiro (1980). The effects of exposure duration on the Bezold-Brücke phenomenon. *Journal of the Color*, 4(4)(19801230), pp. 150–155.

Mollon, J (1983). *Color Vision*. London: Academic Press.

Monnin, D (2003). *Réseaux neuronaux cellulaires et traitement d'images, une approche analytique de la conception d'opérateurs*. Ph.D thesis, Joseph Fourier Univerity, Grenoble, France.

Monnin, D, L Merlat, A Köneke and J Herault (1998). Design of cellular neural networks for binary and gray level image processing, *Proc. of ICANN 98*, Skövde, Sweden, pp. 743–748.

Monnin, D, L Merlat, A Köneke and J Hérault (2000). Straightforward design of robust cellular networks for image processing. *CNNA'2000*, 23–25 MAY 2000, CATANIA, Italy.

Moons, T, L Van Gool, M Van Diest and E Pauwels (1993). Affine reconstruction from perspective image pairs. In *Proceeding of the DARPA-ESPRIT Workshop on Applications of Invariants in Computer Vision*, Azores, Portugal, pp. 249–266.

Morrone, MC and DC Burr (1988). Feature detection in human vision: A phase-dependent energy model. *Proceedings of the Royal Society of London, Series B, Biol. Sci.*, 235(1280), pp. 221–245.

Movshon, JA, EH Adelson, MS Gizzi and WT Newsome (1986). The analysis of moving visual patterns. In *Experimental Brain Research*, Supplementum II: Pattern recognition mechanisms. C Chagas, R Gattass and C Gross (Eds.), Springer, New York, pp. 117–151.

Mullen, KT (1985). Contrast sensitivity of human color vision to red-green and blue-yellow chromatic gratings. *Journal of Physiology*, 359, pp. 381–400. (London)

Nadal, JP and N Parga (1995). Information transmission by networks of non linear neurons. *Proceedings of the Third Workshop on Neural Networks:*

From Biology to High Energy Physics, 1994 (Elba, Italy). *Intl. Journal of Neural Systems*, Supplement 1995, pp. 153–157.

Nagy, AL, RT Eskew and RM Boynton (1987). Analysis of color-matching ellipses in a cone-excitation space. *Journal of the Optical Society of America*, 4, pp. 756–768.

Nakamura, Y, BA McGuire and P Sterling (1980). Interplexiform cell in cat retina: Identification by uptake of [gamma]-[3H] aminobutyric acid and serial reconstruction. *Proceedings of the National Academic Sciences of the USA*, 77, pp. 658–661.

Nalwa, VS (1993). *A Guided Tour of Computer Vision*. Addison-Wesley Publishing Company.

Novak, LG and J Bullier (1997). The timing of information transfer in the visual system. In *Cerebral Cortex*, JH Kaas, K Rockland and A Peters (Eds.), pp. 205–241. New York: Plenum.

O'Bryan, PM (1973). Properties of the depolarizing synaptic potential evoqued by peripheral illumination in cones of the turtle retina. *Journal of Physiology*, 235, pp. 207–223.

Ohki, K, Y Matsuda, A Ajima, DS Kim and S Tanaka (2000). Arrangement of orientation pinwheel centers around area 17/18 transition zone in cat visual cortex. *Cerebral Cortex*, 10(6), pp. 593–601.

Ohzawa, I, Sclar and RD Freeman (1985). Contrast gain control in the cat's visual system, *Journal of Neurophysiology*, 54, pp. 651–667.

Oliva, A (2005). Gist of the Scene. In *Neurobiology of Attention*, L Itti, G Rees and JK Tsotsos (Eds.), pp. 251–256. Elsevier.

Oliva, A and A Torralba (2001). Modeling the shape of the scene: A holistic representation of the spatial envelope. *International Journal of Computer Vision*, 42(3), pp. 145–175.

Osorio, D, DL Ruderman and TW Cronin (1998). Estimation of errors in luminance signals encoded by primate retina resulting from sampling of natural images with red and green cones. *Journal of Optical Society of America A*, 15(1), pp. 16–22.

Osterberg, G (1935). Topography of the layer of rods and cones in the human retina. *Acta Ophthal.* 6(suppl.), pp. 1–103.

Packer, OS and DM Dacey (2001). Receptive field structure of H1 horizontal cells in macaque monkey retina. *Journal of Vision*, 2(4), pp. 272–292.

Parraga, CA (1995). *Spatiochromatic Information Content of Natural Scenes*. M.Sc. thesis, Dept of Psychology, University of Bristol, UK.

Parraga, CA, G Brelstaff, T Troscianko and I Moorehead (1998). Color and luminance information in natural scenes. *J Opt Soc Am A Opt Image*

Sci Vis, 15(3), pp. 563–569. Published erratum appeared in *J Opt Soc Am A Opt Image Sci Vis June 1998*, 15(6).

Parraga, CA, T Troscianko and DJ Tolhurst (2002). Spatiochromatic properties of natural images and human vision. *Current Biology*, 12(6), pp. 483–487.

Pascual-Leone, A and V Walsh (2001). Fast backprojections from the motion to the primary visual area necessary for visual awareness. *Science*, 292, pp. 510–512.

Paulus, W and A Kröger-Paulus (1983). A new concept of retinal color coding. *Vision Research*, 23, pp. 529–540.

Perry, VH and A Cowey (1985). The ganglion cell and cone distributions in the monkey's retina: implications for central magnification factors. *Vision Research*, 25(12), pp. 1795–810.

Petrou, M, R Piroddi and S Chandra (2004). Irregularly sampled scenes. In *Image and Signal Processing for Remote Sensing*, SPIE-5573, Maspalomas, Spain, September 2004.

Pettet, MW and CD Gilbert (1992). Dynamic changes in receptive field size in cat primary visual cortex. *Proceedings of the National Academy Sciences of USA*, 89, pp. 8366–8370.

Pflug, R and R Nelson (1989). Dopaminergic effects on response kinetics of rabbit and cat horizontal cells. *Investigative Ophthalmology and Visual Science*, 30(18) (Suppl).

Pham, DT, P Garrat and C Jutten (1992). Separation of a mixture of independent sources through a maximum likelihood approach. *Proc. EUSIPCO*, pp. 771–774.

Pingault, M and D Pellerin (2004). Motion estimation of transparent objects in the frequency domain, *Signal Processing*, 84(4), pp. 709–719.

Pingault, M, E Bruno and D Pellerin (2003). A robust multiscale B-spline function decomposition for estimating motion transparency. *IEEE Transactions on Image Processing*, 12(11), pp. 1416–1426.

Poggio, GF and B Fischer (1977). Binocular interaction and depth sensitivity in striate and prestriate cortex of behaving rhesus monkey. *Journal of Neurophysiology*, 40, pp. 1392–1405.

Polat, U and D Sagi (1999). Spatial interactions in human vision: From near to far via experience dependent cascades of connections, *Proceedings of the National Academy Sciences of USA*, 91, pp. 1206–1209.

Pollen, DA and SF Ronner (1983). Visual cortical neurons as localized spatial frequency filters. *IEEE Transactions on Systems, Man and Cybernetics*, 13, pp. 907–916.

Potter, MC and BA Faulconer, (1975). Time to understand pictures and words. *Nature*, 253, pp. 437–438.

Prince, SJ, BG Cumming and A.J Parker (2002). Range and mechanism of encoding of horizontal disparity in macaque V1. *Journal of Neurophysiology*, 87(1), pp. 209–221.

Pugh, EN and TD Lamb (1993). Amplification and kinetics of the activation steps in phototransduction. *Biochimica et Biophysica Acta*, 1141, pp. 111–149.

Purdy, D (1931). Spectral hue as a function of intensity. *American Journal of Psychology*, 43, pp. 541–559.

Rabbetts, RB and AG Bennett (2007). *Bennett and Rabbetts' Clinical Visual Optics*. Elsevier Health Science.

Ramachandran, VS, S Cobb and D Rogers-Ramachandran (1988). Perception of 3-D structure from motion: The role of velocity gradients and segmentation boudaries. *Perception and Psychophysics*, 44(4), pp. 390–393.

Reid, RC and JM Alonso (1995). Specificity of monosynaptic connections from thalamus to visual cortex, *Nature*, 378, pp. 281–284.

Ribeiro, E and ER Hancock, (2001). Shape from periodic texture using the eigen vectors of local affine distortion. *IEEE Transactions on Pattern Analysis and Machine Intelligence*, 23(12), pp. 1459–1465.

Ringach, DL (2002). Spatial structure and symmetry of simple-cell receptive fields in macaque primary visual cortex. *Journal of Neurophysiology*, 88(1), pp. 455–63.

Rodieck, RW and J Stone (1965). Response of cat retinal ganglion cells to moving visual patterns. *Journal of Neurophysiology*, 28, pp. 819–832.

Rodieck, RW, KF Binmoeller and J Dineen (1985). Parasol and midget ganglion cells of the human retina. *Journal of Comparative Neurology*. 233, pp. 115–132.

Rojers, AS and EL Schwartz (1990). Design considerations on a space-variant visual sensor with complex logarithmic geometry. *Proc. of the 10th Int. Conference on Pattern Recognition*, 2, pp. 278–285.

Rolls, E.T (1992). Neurophysiological mechanisms underlying face processing within and beyond the temporal cortical visual areas. *Phil. Trans. Roy. Soc. B*, 335, pp. 11–21.

Rolls, ET (1994). Brain mechanisms for invariant visual recognition and learning. *Behavioral Processes*, 33, pp. 113–138.

Rolls, ET and MJ Tovee (1995). Sparseness of the neuronal representation of stimuli in the primate temporal visual cortex. *Journal of Neurophysiology*, 73, pp. 713–726.

Roorda, A and DR Williams (1998). Objective identification of M and L cones in the living human eye. *Investigative Ophtalmology and Visual Science* (ARVO abstracts), 39, S204.

Roorda, A and DR Williams (1999). The arrangement of the three cone classes in the living human eye. *Nature*, 397(6719), pp. 520–522.

Ruderman, DL (1994). The statistics of natural images. *Network*, 5, 517–48.

Rushton, WA (1965). Visual adaptation. *Proc. R. Soc. Lon. B*, 162, pp. 20–46.

Ryer, AD (1997). *Light Measurement Handbook*. Newburyport, MA: International Light, Inc., http://www.Intl-Light.com/handbook/.

Sakai, K and H Finkel (1997). Spatial-frequency analysis in the perception of perspective depth. *Network: Computation in Neural Systems*, 8(3), pp. 335–352.

Sakai, K and LH Finkel (1995). Characterisation of spatial frequency in the perception of shape from texture, *Journal of Optical Society of America A*, 12, pp. 1208–1224.

Sakitt, B and HB Barlow (1982). A model for the economical encoding of the visual image in cerebral cortex. *Biological Cybernetics*, 43(2), pp. 97–108.

Salinas, E and TJ Sejnowski (2001). Gain modulation in the central nervous system: Where behavior and computation meet. *Neurophysiology*, 7(5), pp. 430–440.

Salzman, CD, CM Murasugi, KH Britten and WT Newsome (1992). Microstimulation in visual area MT: Effects on direction discrimination performance, *Journal of Neuroscience*, 12, pp. 2331–2355.

Sammon, JW (1969). A nonlinear mapping algorithm for data structure analysis, *IEEE Transactions on Computer*, CC-18(5), pp. 401–409.

Sarpeshkar, R, J Kramer, G Indiveri and C Koch (1996). Analog VLSI architectures for motion processing: From fundamental limits to system applications. *Proc. IEEE*, 84, pp. 969–987.

Schiller, PH and JG Malpeli (1978). Functional specificity of lateral geniculate nucleus laminae of the rhesus monkey. *Journal of Neurophysiology*, 41, pp. 788–797.

Schiller, PH, BL Finlay and SF Volman (1976). Quantitative studies of single-cell properties in monkey striate cortex. I. Spatiotemporal organization of receptive fields. *Journal of Neurophysiology*, 39, pp. 1288–1319.

Schmolesky, MT, Y Wang, DP Hanes, KG Thompson, S Leutgeb, JD Schall and AG Leventhal (1998). Signal timing across the macaque visual system. *Journal of Neurophysiology*, 79(6), pp. 3272–8.

Schnapf, JL, TW Kraft, BJ Nunn and DA Baylor (1988). Spectral sensitivity of primate photoreceptors. *Visual Neuroscience*, 1, pp. 255–261.

Schwartz, EL (1980). Computational anatomy and functional architecture of striate cortex: A spatial mapping approach to perceptual coding. *Vision Research*, 20, pp. 645–670.

Scott, DW and JR Thompson (1983). Probability density estimation in higher dimensions, In *Proceedings of the Fifteenth Symposium on the Interface*, JR Gentle (Ed.), North-Holland, Elsevier Science Publishers, Amsterdam, New York, Oxford, pp. 173–179.

Sere, B, C Marendaz and J Hérault (2000). Nonhomogeneous resolution of images of natural scenes. *Perception*, 29, pp. 1403–1412.

Shapley, RM and JD Victor (1979). The contrast gain control of the cat retina. *Vision Research*, 19, pp. 431–434.

Shepherd, GM (1994). *Neurobiology*, 3rd (ed.). Oxford: Oxford University Press.

Shi, BE (1998). Gabor-type filtering in space and time with cellular neural networks. *IEEE Transactions on Circuits and Systems-I*, 45(2), pp. 121–132.

Shi, BE, T Roska and LO Chua (1993). Design of linear cellular neural networks for motion sensitive filtering. *IEEE Transactions on Circuits and Systems-II*, 40(5), pp. 320–331.

Shi, BE, T Roska and LO Chua (1998). Hyperacuity in cellular neural networks and the measurement of optical flow. *International Journal of Circuit Theory and its Applications*, 26, pp. 343–364.

Shoham, D, M Hubener, S Schulze, A Grinvald and T Bonhoeffer (1997). Spatio-temporal frequency domains and their relations to cytochrome oxydase staining in cat visual cortex. *Nature*, 385, pp. 529–533.

Shostak, Y, Y Ding and VA Casagrande (2003). Neurochemical comparison of synaptic arrangements of parvocellular, magnocellular and konio-cellular geniculate pathways in Owl monkey visual cortex. *Journal of Comparative Neurology*, 456, pp. 12–28.

Sicard, G (1999). *Une rétine bio-inspirée analogique pour un capteur de vision "intelligent" adaptatif.* Thèse de doctorat de l'Institut National polytechnique de Grenoble.

Silveira, LCL, U Grünert, J Kremers, BB Lee and PR Martin (2005). Comparative Anatomy and Physiology of the Primate Retina. In *The Primate Visual System: A Comparative Approach*, J. Kremers (Ed.), John Wiley and Sons.

Silverman, BW (1986). *Density Estimation for Statistics and Data Analysis*. Chapman and Hall.

Simoncelli, EP (2003). Vision and statistics of the visual environment. *Current Opinion in Neurobiology*, 13, pp. 144–149.

Simoncelli, EP and BA Olshausen (2001). Natural image statistics and neural representation. *Annual Review of Neuroscience*, 24, pp. 1193–1216.

Smirnakis, SM, MJ Berry, DK Warland, W Bialek and M Meister (1997). Adaptation of retinal processing to image contrast and spatial scale. *Nature*, 386, pp. 69–73.

Smith, AT, W Curran and OJ Braddick (1999). What motion distributions yield global transparency and spatial segmentation? *Vision Research*, 39, pp. 1121–1132.

Smith, VC and J Pokorny (1975). Spectral sensitivity of the foveal cone photopigments between 400 and 500 nm. *Vision Research*, 15, pp. 161–171.

Smith, WJ (1966). *Modern Optical Engineering*. New York: McGraw Hill.

Smithson, H and Q Zaidi (2004). Color constancy in context: Roles for local adaptation and levels of reference. *Journal of Vision*, 4, pp. 693–710.

Smithson, HE and JD Mollon (2004). Is the S-opponent chromatic subsystem sluggish? *Vision Research*, 44, pp. 2919 –2929.

Spillman, L and JS Werner (1990). *Visual Perception: The Neurophysiological Foundations*. Academic Press.

Spinei, A, D Pellerin and J Hérault (1998). Spatio-temporal energy-based method for velocity estimation. *Signal processing*, 65(3), pp. 347–362.

Spinei, A and D Pellerin (2001). Motion estimation of opaque or transparent objects using triads of Gabor filters. *Signal Processing*, 81(4), pp. 845–853.

Stemmler, M and C Koch (1999). How voltage-dependent conductances can adapt to maximize the information encoded by neuronal firing rate, *Nature Neuroscience*, 2, pp. 521–527.

Stiles, WS (1959). Color vision: The approach through increment threshold sensitivity. *Procceedings of the National Acadamy of Science*, 75, pp. 100–114.

Stimson, A (1974). *Photometry and Radiometry for Engineers*. New York: John Wiley and Sons.

Super, BJ and AC Bovik (1995a). Shape from texture using local spectral moments. *IEEE Transactions on Pattern Analysis and Machine Intelligence*, 17(4), pp. 333–343.

Super, BJ and AC Bovik (1995b). Planar surface orientation from texture spatial frequencies. *Pattern Recognition*, 28(5), pp. 728–743.

Swindale, NV, JA Matsubara and MS Cynader (1987). Surface organization of orientation and direction selectivity in cat area. *Journal of Neuroscience*, 18(7), pp. 1414–1427.

Szmajda, BA, U Grünert and PR Martin (2005). Mosaic proper properties of midget and parasol ganglion cells in the marmoset retina. *Visual Neuroscience*, 22, pp. 395–404.

Talbot, SA and WH Marshall (1941). Physiological studies on neural mechanisms of visual localization and discrimination. *American Journal of Ophtalmology*, 24, pp. 1255–1263.

Tanner, J and C Mead (1988). An integrated analog optical motion sensor. In *VLSI Signal Processing, Volume 2*, RW Brodersen and HS Moscovitz (Eds.), pp. 59–87. New York: IEEE.

Teissier, P, A Guérin-Dugué and JL.Schwartz (1998). Models for audiovisual fusion in a noisy-vowel recognition task, *Journal of VLSI Signal Processing*, 20, pp. 25–44.

Teranishi, T, K Negishi and S Kato (1983). Dopamine modulates S-potential amplitude and dye-coupling between external horizontal cells in carp retina. *Nature*, 301, pp. 243–246.

Teufel, HJ and C Wehrhahn (2004). Chromatic induction in humans: How are the cone signals combined to provide opponent processing? *Vision Research*, 44, pp. 2425 –2435.

Thorpe, S, D Fize and C Marlot (1996). Speed of processing in the human visual system. *Nature*, 381(6582), pp. 520–522.

Tistarelli, M and G Sandini (1993). On the advantages of polar and logpolar mapping f for direct estimation o of time-to-impact from optical flow. *IEEE Transactions on Pattern Analysis and Machine Intelligence*, 1(14), pp. 401–410.

Tootell, RB, MS Silverman and RL De Valois (1981). Spatial frequency columns in primary visual cortex. *Science*, 214(4522), pp. 813–815.

Torgerson, WS (1952). Multidimensional scaling, I: Theory and method. *Psychometrika*, 17, pp. 401–419.

Torralba, A and A Oliva (2003). Statistics of natural image categories. *Network: Computing Neural Systems*, 14, pp. 391–412.

Torralba, AB (1999). *Analogue Architectures for Vision: Cellular Neural Networks and Neuromorphic Circuits*. Ph.D thesis, INPG, Grenoble.

Torralba, AB and J Hérault (1999a). An efficient neuromorphic analog network for motion estimation. *IEEE Transactions on Circuits and Systems-I: Special Issue on Bio-Inspired Processors and CNNs for Vision*, 46(2), pp. 269–280.

Torralba, AB and J Hérault (1999b). Asymmetrical filters for vision chips: A basis for the design of large sets of spatial and spatio-temporal filters. *7th Int. Conf on Microelectronics for Neural, Fuzzy and Bio-inspired Systems, Microneuro'99*. IEEE Computer Society PR 00043, pp. 224–231.

Torralba, AB, D Alleysson and J Hérault (1998). Spatio-chromatic processing in the human retina: towards an optimal trade-off between spatial resolution of luminance and range color perception. *European Conf. on Visual Perception*, Oxford, UK.

Torre, V, HR Matthews and V Lamb (1986). Role of calcium in regulating the cyclic GMP cascade of phototransduction in retinal rods. *Proc. Natl. Acad. Sci*, 83, pp. 7109–7113.

Trussell, H and RE Hartwig (2002). Mathematics for demosaicing. *IEEE Transanctions on Image Processing*, 11(4), pp. 492–495.

Tsutsui, KI, H Sakata, T Naganuma and M Taira (2002). Neural correlates for perception of 3D surface orientation from texture gradient. *Science*, 298(5592), pp. 409–412.

Turiel, A and N Parga (2000). Multiscaling and information content of natural color images. *Physical Review*, 62(1), pp. 1138–1148.

Valeton, J M and D Van Norren, (1983). Light adaptation of primate cones: An analysis based on extracellular data. *Vision Research*, 23(12), pp. 1539–1547.

van der Schaaf, A and JH van Hateren (1996). Modeling of the power spectra of natural images: Statistics and information. *Vision Research*, 36, pp. 2759–2770.

van Hateren, H (2005). A cellular and molecular model of response kinetics and adaptation in primate cones and horizontal cells. *Journal of Vision*, pp. 331–347.

Venkataraman, V, TV Duda, KW Koch and RK Sharma (2003). Calcium-modulated guanylate cyclase transduction machinery in the photoreceptor-bipolar synaptic region. *Biochemistry*, 42, pp. 5640–5648.

Verri, A, F Girosi and V Torre (1990). Differential techniques for optical flow. *Journal of Optical Society of America A*, 7, pp. 912–922.

Verweij, J, M Kamermans and H Spekreise (1996). Horizontal cells feedback to cones by shifting the cone calcium-current activation range. *Vision Research*, 36, pp. 3943–3953.

Victor, JD (1987). The dynamics of the cat retinal X cell center. *Journal of Physiology*, 386, pp. 219–246.

Vigneron V, V Maiorov, R Berndt, JJ Sanz-Ortega and P Schillebeeckx (1997). Neural network application to enrichment measurements with NAI detectors. *VCCSR Proceedings*, Vienna, November 1997.

Viollet, S and N Franceschini (1999), Visual servo system based on a biologically-inspired scanning Sensor, In *Decentralised control and Sensory Fusion*, P Schenker and R MacKee (Eds.), S.P.I.E., Bellingham, U.S.A.

Virsu, V and R Näsänen (1978). Cortical magnification factor predicts the photopic contrast sensitivity of peripheral vision. *Nature*, 271, pp. 54–56. (London)

von Bezold, W (1874). *Die Farbenlehre in Hinblick auf Kunst und Kunstgewerbe*. Braunschweig.

von Blanckensee, HT (1981). *Spatio-temporal Properties of Cells in Monkey Lateral Geniculate Nucleus*. Doctoral dissertation, University of California, Berkeley.

von der Heydt, R and E Peterhans (1989). Mechanisms of contour perception in monkey visual cortex. I. Lines of pattern discontinuity. *Journal of Neuroscience*, 9, pp. 1731–1748.

von Kries, J (1902). Chromatic Adaptation, Festschrift der Albrecht-Ludwig-Universität, Freiburg. Translation. D. L. MacAdam, *Sources of Color Vision*, MIT Press, Cambridge (1970), pp. 109–119.

Vos, JJ (1986). Are unique and invariant hues coupled? *Vision Research*, 26, pp. 337–342.

Waessle, H and BB Boycott (1991). Functional architecture of the mammalian retina. *Physiological Reviews*, 71(2), pp. 447–480.

Wallis, G (2001). Linear models of simple cells: Correspondence to real cell responses and space spanning properties. *Spatial Vision*, 14(3,4), pp. 237–260.

Walraven, J, C Enroth-Cugell, DC Hood, DAI Macleod and JL Schnapf (1990). The control of visual sensitivity: Receptoral and postreceptoral processes. In *Visual Perception: The Neurophysiological Foundations*, L Spillmann and JS Werner (Eds.), pp. 53–101. San Diego, CA: Academic.

Wandell, BA (1995). *Foundations of Vision*. Sunderland Mass: Sinauer Press.

Wässle, H, U Grünert, J Röhrenbeck and BB Boycott (1990). Retinal ganglion cell density and cortical magnification factor in the primate. *Vision Research*, 30(11), pp. 1897–1911.

Watson, AB and AJ Ahumada (1983). A look at motion in the frequency domain. In *Motion: Perception and Representation*, JK Tsotsos (Ed.), pp. 1–10. New York, Association for computing machinery.

Webster, MA and JD Mollon (1995). Color constancy influenced by contrast adaptation. *Nature*, 373, pp. 694–698.

Weliky, M, WH Bosking and D Fitzpatrick (1996). A systematic map of direction preference in primary visual cortex. *Nature*, 379, pp. 725–728.

Werblin, F (1991) Synaptic connections, receptive fields, and patterns of activity in the tiger salamander retina. *Investigative Ophthalmology and Visual Science*, 32, pp. 459–483.

Werblin, F, G Maguire, P Lukasiewicz, S Eliasof and SM Wu (1988). Neural interaction mediating the detection of motion in the retina of tiger salamander. *Visual Neuroscience*, 1, pp. 317–329.

Werblin, FS and JE Dowling (1969). Organization of the retina of the mudpuppy, Necturus maculosus. II. Intracellular recording. *Journal of Neurophysiology*, 32, pp. 339–355.

Westheimer, G (2006). Specifying and controlling the optical image on the human retina. *Progress in Retinal and Eye Research*, 25, pp. 19–42.

Williams, DR (1988). Topography of the foveal cone mosaic in the living human eye. *Vision Research*, 28, pp. 433–454.

Witkovsky, P, S Stone and D Tranchina (1989). Photoreceptor to horizontal cell synaptic transfer in the xenopus retina: Modulation by dopamine ligands and a circuit model for interactions of rod and cone inputs. *Journal of Neurophysiology*, 62(4), pp. 864–881.

Wohrer, A and P Kornprobst (2008). Virtual retina: A biological retina model and simulator, with contrast gain control. *Journal of Computational Neuroscience*, DOI 10.1007/s10827-008-0108-4.

Wohrer, A, P Kornprobst and T Vieville (2006). From Light to Spikes: a Large-Scale Retina Simulator. *International Joint Conference on Neural Networks IJCNN '06*, pp. 4562–4570.

Worgotter, F (1999). An ASIC-chip for stereoscopic depth analysis in video-real-time based on visual cortical cell behavior. *Int J Neural Syst*, 9(5), pp. 417–422.

Wyszecki, G and GH Fielder (1971). New color-matching ellipses. *Journal of the Optical Society of America*, 61, pp. 1135–1152.

Wyszeki, G and WS Stiles (1982). *Color Science: Concepts and Methods, Quantitative Data and Formulae*. New York: Wiley.

Xu, X, J Ichida, Y Shostak, AB Bonds and VA Casagrande (2002). Are primate lateral geniculate nucleus cells really sensitive to orientation or direction? *Visual Neuroscience*, 19, pp. 97–108.

Yang, XL and SM Wu (1991). Feedforward lateral inhibition in retinal bipolar cells: Input-output relation of the horizontal cell-depolarising bipolar cell synapse. *Proceedings of the National Academy of Sciences of USA*, 88, pp. 3310–3313.

Yellott, Jr. JI (1983). Spectral consequences of photoreceptor sampling in the rhesus retina. *Science*, 221(4608), pp. 382–385.

Yu, YG and TS Lee (2005). Adaptive contrast gain control and information maximization. *Neurocomputing*, 65–66, pp. 111–116.

Zipser, K, VAF Lamme and PH Schiller (1996). Contextual modulation in primary visual cortex. *Journal of Neurosciences*, 16(22), pp. 7376–7389.

Index